Man and Environment Series
Editor: David Burtenshaw

Climate and Society

Allen Perry
and Vivien Perry

First published in 1986 by
Bell & Hyman Limited Denmark House
37-39 Queen Elizabeth Street London SE1 2QB

British Library Cataloguing in Publication Data

Perry, Allen H.
Climate and society.—(Man and environment; v. 6)
1. Climate—Social aspects
I. Title II. Perry, Vivien C. III. Series
304.2'5 GF71

ISBN 0 7135 2498 7

Typeset by August Filmsetting, Haydock, St. Helens
Printed in Great Britain by R.J. Acford Ltd, Chichester

Contents

Acknowledgements

For permission to reproduce copyright material the authors and publisher are indebted to: Academic Press for a diagram from *Biometeorological Methods* by R.E. Munn (Fig. 4.4); Blackie and Son Ltd for a diagram from *Physical Geography* by D. Wilcock (original source R.G. Barry and R.J. Chorley) (Fig. 1.9); Collins Publishers for a diagram from *The Irish Landscape* by F. Mitchell (Fig. 3.4); Department of the Environment, Warren Springs Laboratory, © Crown Copyright (Figs 9.7, 9.8): The Geographical Association for a diagram from *Geography*, 46 (1961) (Fig. 7.6); Mary Glasgow Publications Ltd, London, for a diagram from *Geo* Series 1, Issue 4, p. 10 (Fig. 3.5); Hamish Hamilton Ltd (Table 2.1); the Controller of Her Majesty's Stationery Office for a diagram from *Services to Agriculture* (Fig. 6.4); Hodder & Stoughton Educational Ltd for a diagram from *Earth, the Living Planet* by Bradshaw, Abbot and Gelethorpe (Fig. 7.5); The Institute of British Geographers for a diagram by M.L. Parry from *Transactions*, 1975 (Fig. 3.11); W.W. Kellogg and *Science* for three diagrams, Figs 1, 2, 4, pp. 1164, 1165, 1169, from *Climate Stabilization: For better or for worse* by W.W. Kellogg and S.H. Schneider, Vol. 186, 27 December 1974. Copyright 1974 by the American Association for the Advancement of Science (Figs 1.3, 1.5, 12.1); H.H. Lamb (Figs 3.9, 9.2); J.G. Lockwood (Fig. 1.7); Longman Group for a diagram from *Environment, Man and Economic Change,* eds A.D.M. Phillips and B.J. Turton (Fig. 1.11); Macmillan Publishing Company for a diagram from the *U.S. Air Force Handbook of Geophysics*, Revised (New York: Macmillan, 1960) (Fig. 4.1); for a diagram from *Environmental Change* by A.S. Goudie, published by Oxford Univeristy Press, © A. Goudie 1983 (Fig. 3.1); Alec Paul and David & Charles Publishers for a diagram which originally appeared in *Weather Forecasting for Agriculture and Industry*, ed. James A. Taylor (Fig. 6.3); Relief and Development Institute (Fig. 5.2); Royal Meteorological Society for a diagram from *Weather*, 34 (1979) (Fig. 4.5) and a table from *Weather*, 13, (1958) (Table 1.1); *Scientific American* and W.H. Freeman and Company (Fig. 7.2); *Sunday Express* Magazine (Figs 9.4, 9.5); Swets and Zeitlinger bv for two diagrams from the *International Journal of Biometeorology* 11 (1), (1967) (Figs 2.2, 2.3); John Wiley & Sons Ltd for two diagrams from *Introduction to Climatology for the Tropics* by J.O. Ayoade (Figs 2.5, 3.3).

Picture credits: Barnaby's Picture Library (Figs 5.1(a) and (b), 6.7(a) and (d), 7.1); Friends of the Earth Trust (Fig. 9.3); GeoScience Features (Figs 1.13, 2.9, 2.13, 2.14); Greater London Council (Fig. 5.4); aerial photography by Handford Photography, Croydon (Fig 7.4); the Controller of Her Majesty's Stationery Office, © Crown Copyright (Figs 5.7, 10.1, 10.2); International Coffee Association (Fig 6.7(b)); International Distillers and Vintners (UK) Ltd (Fig. 6.7(c)); Milton Keynes Development Corporation (Fig. 11.2); OXFAM (Fig. 3.10); Popperfoto (Fig. 5.3); the Royal Meteorological Society and the Royal Yachting Association (Fig. 10.4); Charles Tait Photographic (Fig. 11.1); Warner Holidays Ltd (Fig. 6.2).

Every effort has been made to trace owners of copyright material, but in some cases this has not proved possible. The publisher would be glad to hear from any further copyright owners not acknowledged here.

Series preface

Alvin Toffler, in his book *Future Shock*, examined the effects of the increasing pace of change in modern society. The speed of change in modern geographical education to the 16–19 age group is no exception. The methodology of the Quantitative Revolution is commonplace and examination boards have now begun to encompass the more challenging framework of the People-Environment approach and the introduction of value-based learning. In addition, the nature of the traditional sixth form has been changed so that the traditional Advanced level student is not the sole type of student. Sixth form and tertiary colleges teaching large groups and the increasing numbers of students in the 'New Sixth' studying for BEC and similar awards, are common today. In addition, only a small proportion of geographers at 16–19 actually proceed to take the subject in further education.

These contrasting demands have influenced the thinking behind this series of short texts aimed at the modern lower sixth student, ninety per cent of whom will not pursue geography beyond the sixth form. We have assumed that the readers will have taken either 'O' level or CSE geography examinations although the content should present no difficulties for the more mature reader. In all the texts it is our intention that the author's enthusiasm for the topic, combined with the interest generated by the presentation of the text, will assist a wide range of learners in and out of the classroom.

Every text can be studied in isolation, which should suit the individual curriculum of any group. Therefore, there is little attempt to cross reference texts which might be used in different order by schools and colleges. In addition, a self-standing set of texts leaves topic selection within a syllabus to the teacher's discretion. Nevertheless, the texts will develop new views of the more traditional sections of geography and introduce others where available material at this level is limited.

All the authors have a common concern for the improvement of geography teaching to the 16–19 group as experienced teachers, or lecturers, or examiners. All try to build on existing knowledge of our environment and to stress the dynamic nature of the environment. Where possible, they have asked the reader to express attitudes and explain values towards the major issues which affect our social, economic and physical environment. Above all, the authors are aware of geography's continuing role in teaching people how to think.

David Burtenshaw
Waterlooville 1983

Foreword

Climate affects all of us in one way or another, yet climatology is often the most neglected aspect of the geography syllabus. When the subject is taught at all little attempt is usually made to highlight the importance of climatic resources to mankind. *Climate and Society* looks at a key aspect of the man–environment relationship that has been neglected for too long. It aims to be stimulating and to provide a text in applied climatology which emphasises the various linkages that exist between mankind and the atmospheric environment.

The book is divided into four major sections, each comprising three chapters. Firstly, the spatial and temporal dimensions of the climate system are explored. Then the impact of the climate system on human beings and their activities, including lesser known topics like econoclimate, is discussed. In the third section, man's influence on the climate system is examined at different scales, ranging from the individual city to the entire planet. In conclusion, the exploitation of the climate system, including pressing issues such as acid rain and desertification, is analysed. A range of study activities, projects and questions are found in each chapter to stimulate students into considering the topics covered and to aid their understanding of the material presented.

Physical geography is now much more people-orientated than it has been in the past and we have sought to contribute to this trend by looking at the 'human' interpretation of the atmosphere. It is hoped the reader will see the need for, and the practical value of, weather studies. We have no doubt that studies of weather in relation to human affairs will gain a progressively more secure place in the syllabus and hope that this book will help to correct the unfortunately prevailing view that climatic studies are dull and peripheral to geography.

1

All systems go!

What is a system?

Everyone reading this book probably uses the word **system** very frequently. We speak of the electoral system, the transport system, the central-heating system and many other kinds of system. When we use these terms, we normally mean that each of these examples consists of interrelated parts that are in an organised relationship with each other and operate together as a complex whole. The systems concept allows us to examine the immensely complex real world and helps us to understand it.

We can divide systems into three main functional categories:

1 **isolated systems** which have boundaries that are closed to the import and export of both mass and energy. These are rare in the real world;
2 **closed systems** in which there is no exchange of matter between the system and its environment but in which there is energy exchange;
3 **open systems** where there is a free exchange of both energy and materials.

The atmosphere, oceans and land surfaces may be thought of as consisting of a chain of open systems which are linked by a cascade of mass and energy. The **output** of one subsystem becomes the **input** for the next subsystem, and this dynamic linkage creates a **cascading system**. The principal cascade in the atmosphere is the **solar energy cascade**, of which the driving force is the sun, which generates energy that passes through the layers of the atmosphere to the earth's surface, and into its surface layers.

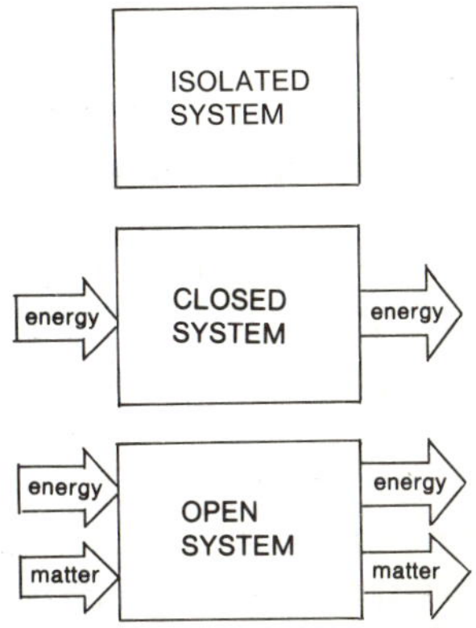

Figure 1.1 Isolated, closed and open energy systems

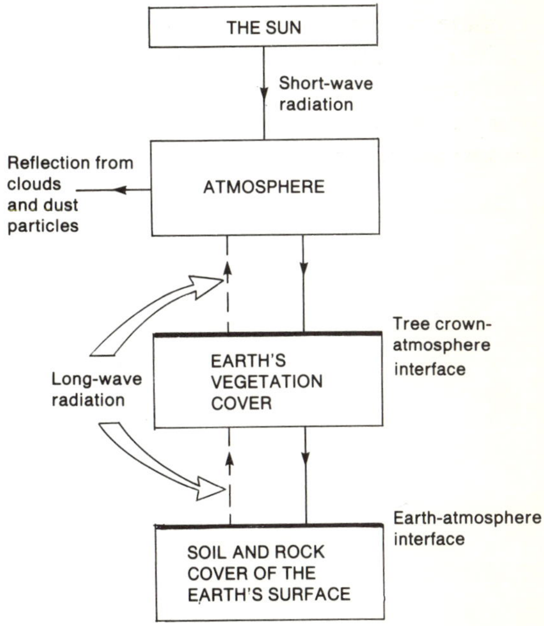

Figure 1.2 An example of an open system

> **Study activities**
>
> 1 Figure 1.2 is an example of an open system. Why do you think it is defined as an open system? List the inputs and outputs of this system.
>
> 2 Construct a diagram to represent a system with which you are familiar. Label the inputs and the outputs.
>
> 3 Give an example of each of the different types of system: isolated; closed; open. Explain why you have chosen each example.

The components of the climate system

The climate system consists of the properties and processes that are responsible for climate. The properties are: **thermal properties** – the temperature of the air, water, ice and land; **kinetic properties** – wind and ocean currents, together with associated vertical movement, and ice movement; **aqueous properties** – the moisture (humidity) of the air, cloud water content, lake levels, ground-water, ice and snow water content; **static properties** – pressure, density of both atmosphere and ocean, the oceanic salinity, the composition of the air, and the physical constants of the system. These properties are linked through the physical processes of the system.

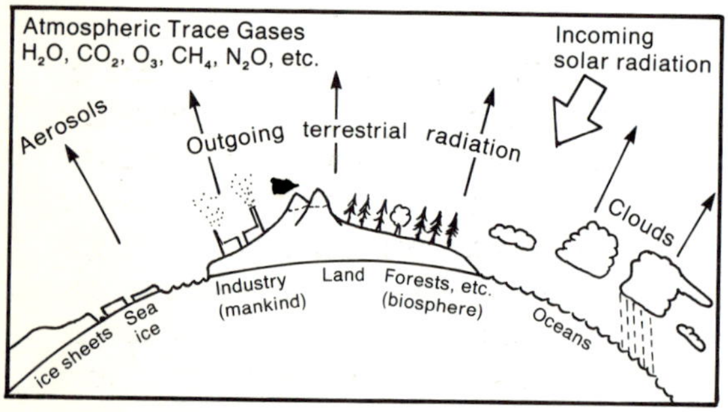

Figure 1.3 Components of the climate system

Figure 1.3 provides a simple representation of the components of the climate system and the relation they bear to one another. The components can be classified into five broad categories:

1 **Atmosphere** This is the gaseous layer that envelops the earth, and is the most variable part of the system. It is through the atmosphere that heat is transferred vertically and horizontally.

2 **Hydrosphere** The hydrosphere is composed of the liquid water over the surface of the earth. It includes the oceans, seas, lakes, rivers and ground-water. In climatic terms, the oceans are most important because they absorb much of the solar radiation reaching the earth's surface and provide an enormous reservoir of heat.

3 **Cryosphere** The world's ice masses and snow deposits, including the continental ice sheets, sea ice, snow cover, lake and river ice, are the constituents of the cryosphere. Significant changes in snow cover occur seasonally but changes in the amount and extent of ice occur at a very slow rate.

4 **Lithosphere** This is the solid crust of the earth together with the upper solid zone of the **mantle**, immediately beneath. It includes the mountain and ocean basins, together with surface rock, sediment and soil. The rate of change is so slow that the lithosphere can be regarded as the most stable component of the system.

5 **Biosphere** The biosphere encompasses that part of the earth that contains life. The basic processes within the biosphere are photosynthesis and respiration, and involve the cycling of hydrogen, oxygen and carbon between solid, liquid and gaseous forms of matter. The biosphere exists at the junction of the lithosphere, the hydrosphere and the atmosphere and occupies only a narrow vertical zone. Man is a principal component of the biosphere, but vegetation also plays a significant role, especially where there is extensive forest cover.

Table 1.1 gives the properties of the earth's surface.

> **Study activity**
>
> The properties given in Table 1.1 can be applied at all scales of the climate system from a large region to a very small area.

Table 1.1 The types and properties of the earth's surface that influence the climate system

Factor	Nature of the factor
Type of surface	
Rock	Type, colour, thermal conductivity.
Soil	Type, texture, colour, air and moisture content, thermal conductivity.
Water	Surface area, depth, movements.
Vegetation	Type, height, density, colour, seasonal change.
Agricultural	Fallow land; type, height and colour of crops; seasònal change.
Urban and industrial	Material (concrete, tarmac, wood, metal, etc.), colour, thermal conductivity; sources of heat, moisture, pollution.
Properties of surface	
Geometrical shape	Flat, convex, concave.
Energy supply	Latitude and altitude, degree of screening of natural horizon, aspect, slope, exposure.
Exposure	Shelter provided by large and small scale relief features; shelter provided by buildings, trees.
Topographic roughness	(a) Rural areas: extent of woodland, grassland, arable; location of wind-breaks and hedges; degree of agglomeration or dispersion of individual buildings. (b) Urban areas: distribution and average height of different types of built-up zones; orientation and exposure of streets, blocks and individual buildings; density of parks, gardens and other open spaces; vertical profiles across area.
Albedo (rate of reflection)	Type of surface.
Radiating capacity	Type and maximum temperature of surface; observed earth radiation.

(after Stringer)

Select four areas: a small rural area; a small urban area (or part of a larger urban area); your garden or a local park; the area around the school. For each area, describe the factors according to the list in the right-hand column of the table. Then attempt to explain the effect that the surface will have on the surrounding climate.

Study questions

1 Can you define: geometric shape; topographic roughness; radiating activity.
2 Can you distinguish kinetic and static properties, positive and negative feedbacks, short and long wave radiation?

Feedback loops

The application of systems theory to climatology has changed its nature completely in recent years. Modern climatology is concerned with the interactions and exchanges that occur between the components of the climate system.

A change in any part of the climate system will affect all other parts of the system. Thus climatic variations can be caused by changes in, for example, the cryosphere or the biosphere. We use particular terms to describe the process of change between the components of the system:

1 **Negative feedback loops** have the effect of maintaining and regulating the stability of the system, rather like a thermostat in a heating system. The atmosphere's driving force is the **Equator–pole temperature gradient**. If this increases, the vigour of the circulation also increases. As the storm systems become larger and stronger they transport additional heat polewards. This is the negative feedback, lessening the increase of temperature difference from Equator to pole, thus assisting in the maintenance of an overall balance of planetary radiation.
2 **Positive feedback loops** are said to occur when an initial change is supported by subsequent changes and the former stability of the

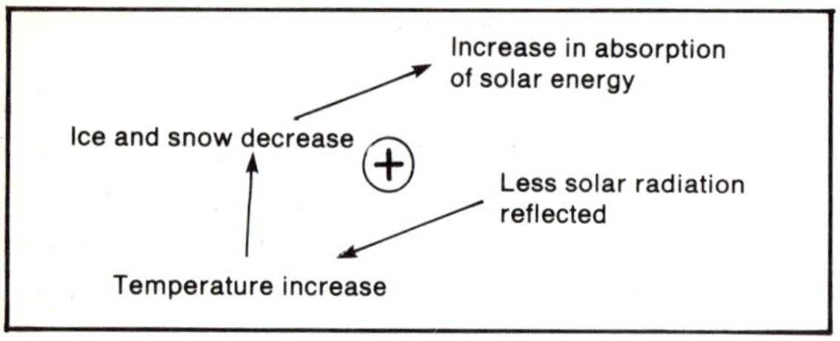

Figure 1.4 A positive feedback loop

system is lost. With positive feedback there is a cumulative effect, reinforcing a particular direction of change. Figure 1.4 illustrates a positive feedback loop. An increase in temperature leads to a decrease in the extent of the ice and snow. This results in an increase in the earth's ability to absorb solar energy and so further heating takes place. Thus change is transmitted through the system.

The concept of a feedback process is very important in the study of climatic variations, and many feedback loops have been identified. However, interactions among many feedback loops that are occurring simultaneously are not yet understood. A specialist committee, set up in 1976 to study climate and climatic variations (Rockefeller Foundation, 1976), reported:

The atmospheric system includes large numbers of both positive and negative feedback loops, and it is often difficult to separate causes and effects. For example, the wind can drive an ocean current which alters the distribution of sea surface temperatures, influencing the atmospheric circulation which determines the strength and the direction of the wind.

The importance of the role of feedback has thus been recognised, but so far mainly at the descriptive level. Nevertheless, the understanding already achieved has been of considerable assistance in improving knowledge on the interactions of the components of the climate system. This is particularly true with the recognition of the fact that changes within components can have repercussions on the whole system. Figure 1.5 illustrates some of the major cause-and-effect, or feedback, linkages.

In order to determine more clearly some specific feedback relationships, a quantitative approach is required. A number of models have been formulated, but their complexity is an indication of the difficulties of unravelling the causes and effects of climatic behaviour.

One value of attempting this process is for the purpose of prediction. Having produced a model, it should then be possible to feed weather observation data into the model and discover what sequence of changes might occur as a result of the initial change. For example,

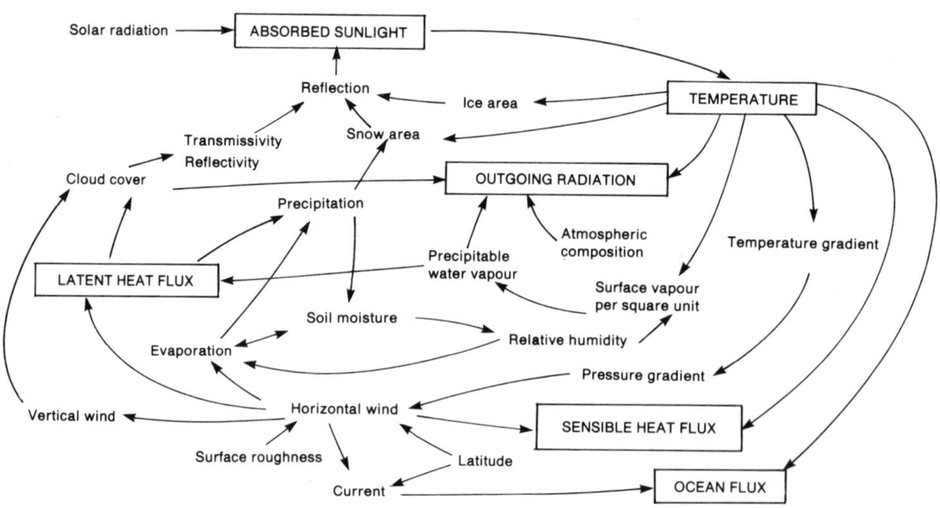

Figure 1.5 Climatic cause-and-effect feedback linkages

measurements of a change in sea temperature in a particular locality can provide the input for the model in order to identify expected weather changes. Quite a lot of work has already been done in an attempt to recognise how different patterns of sea temperature produce specific climatic features, e.g. the spell of cold winters in Europe during the 1960s, and the snow cover of the United States of America during the early spring of 1973.

Study activity

Study Figure 1.5. It appears very complex, but it can be used as a resource for identifying feedback loops. The following diagram shows one of the feedbacks from Figure 1.5.

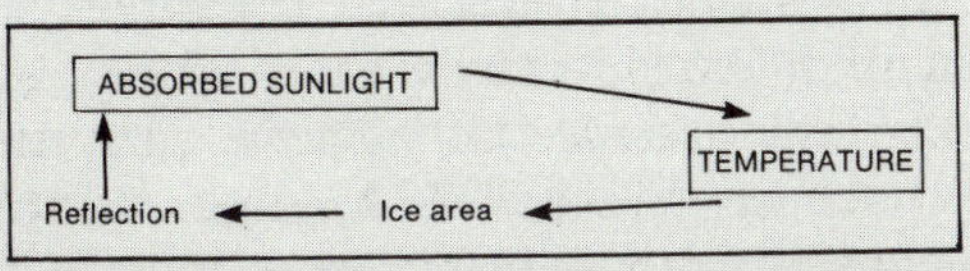

Draw diagrams of the further feedback loops from temperature through reflection to absorbed sunlight.

Study question

What types of weather would it be most important economically and/or socially to be able to predict well in advance for the British Isles? Why?

Flows within the system

The flows that occur within a system may be considered as the circulation of energy and materials through the system. Within the climate system, imports of energy or material are not necessarily exported in the same form. Also these inputs and outputs do not always balance in the short term because both energy and materials can be stored within the system.

The two principal flows or circulations within the climate system relate to heat and to water, termed respectively the **energy budget** and the **hydrological cycle**.

The energy budget

The source of energy imported into the system is the sun, which supplies energy in the form of short-wave radiation. Short-wave radiation cannot normally be absorbed by gases but only by liquid or solid substances. Thus the air itself does not receive energy directly from **insolation** (solar radiation), and only a small proportion of the short-wave radiation received is productive. Most natural systems are inefficient with a considerable 'wastage'. The same is true of the climate system.

Of the total amount of radiation received, a third is lost by being reflected back into space at the edge of the outer atmosphere. Of the re-

Figure 1.6 The vertical energy budget

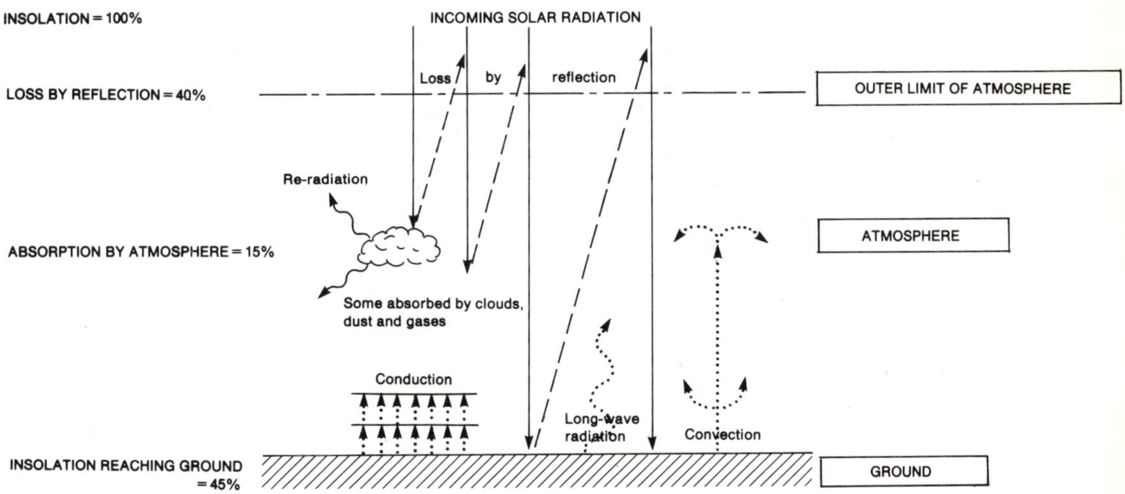

mainder, about one-third is absorbed and scattered by dust, clouds and water vapour in the atmosphere before reaching the ground.

Approximately one-third of the radiation that reaches the earth's surface is reflected back into space, but the amount varies enormously according to the type of surface. The rate of reflection is called the **albedo** and surfaces which influence the albedo include snow or ice cover, water cover, forest, grassland, desert, and urban areas.

There are two main methods of measuring the albedo. Traditionally, an approximate percentage was calculated for each type of surface, as shown in Table 1.2.

More modern methods involve the measurement of the degree of brightness as recorded on satellite photographs. Examination of a satellite photograph will show that ice is very bright, while dense forest, like the Amazonian forest, is very dark. Those are the two extremes, with the remaining surfaces between, in approximately the same ratios as in Table 1.2.

Table 1.2 The albedo for different surfaces

Surface	Albedo (as percentage of incoming solar radiation that is reflected)
Fresh, dry, snow	80–95
Sea ice	30–40
Dry, light, sandy soils, e.g. deserts	35–45
Meadows	15–25
Dry steppe grassland	20–30
Coniferous forest	10–15
Deciduous forest	15–20

Having examined the primary source of energy in the heat budget, and the variations that occur owing to heat loss in the atmosphere and with different surfaces, we can look at the average radiation received over the earth's surface, shown in Figure 1.7. Amounts decrease polewards from the Equator because the inclination and the density of radiation is greatest close to the Equator and least close to the poles.

Figure 1.7 Average annual global radiation

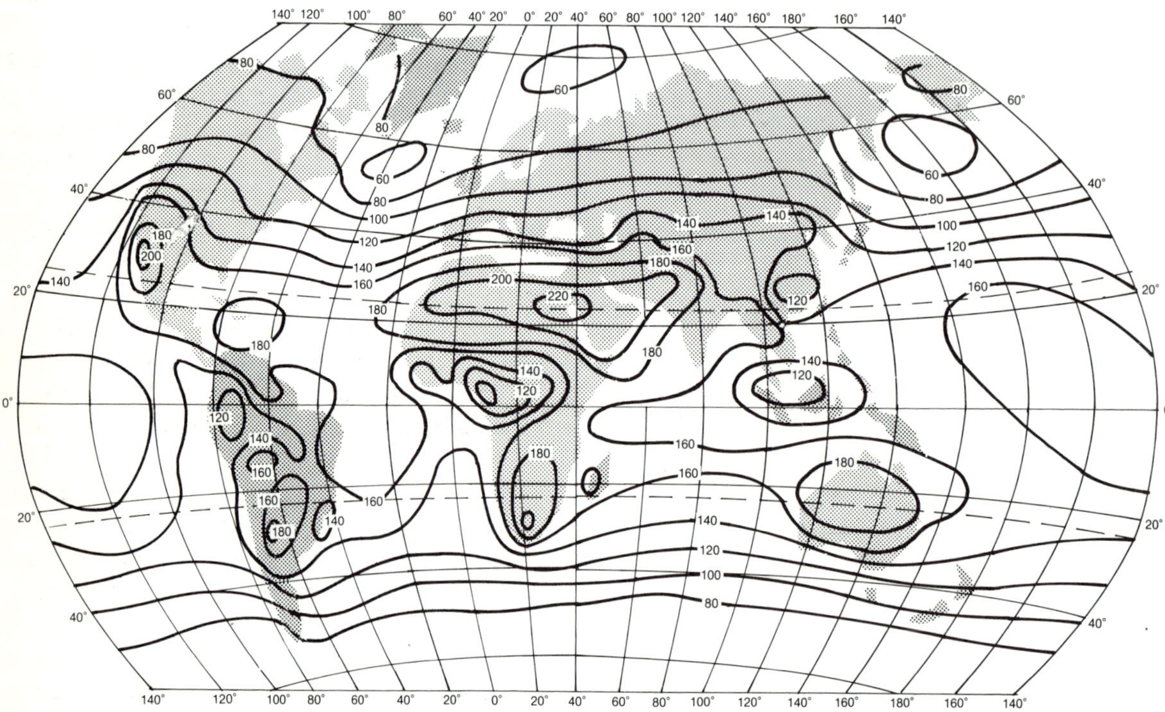

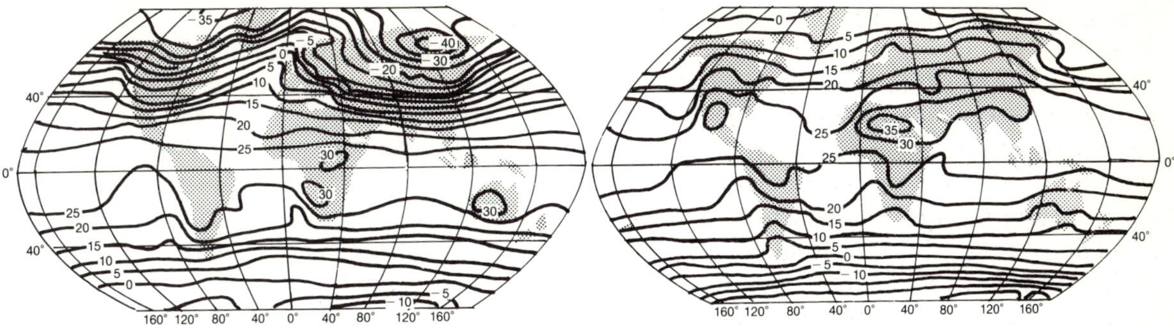

Figure 1.8 *Mean sea-level temperatures in January (left) and July (right)*

Broadly, land masses in the tropical and subtropical regions receive more radiation than the water surfaces.

However, if Figure 1.7 is compared with the map of mean temperatures, shown in Figure 1.8, it can be seen that there are a number of differences. These are largely a result of the effects of the release of heat from the earth's surface. In January, the temperatures of the northern hemisphere's sea areas are higher than the adjacent land masses, therefore producing a very different pattern from the receipt of radiation. However, the July distribution relates more closely to the radiation map, with the maximum temperatures occurring in the same regions as the maximum radiation.

Study questions

1 Describe the anomalies in the pattern of radiation shown in Figure 1.7. On a uniform surface, there would be a uniform decrease of temperature poleward. However, the existence of contrasting surfaces interferes with this pattern. The relative positions of land and sea produce distortions in the theoretical latitudinal distribution. These distortions are termed **anomalies**, so in order to answer this question, you must look for patterns of **isolines** (lines joining together equal values) which do not correspond to a uniform pattern of decrease from Equator to pole.

2 Locate the main anomalies in the temperature pattern for: (a) January, and (b) July (Figure 1.8). Attempt to account for these under the headings: (a) sea/land surfaces, and (b) types of land surface.

Once the earth has received and absorbed radiation from the sun, it is then able to radiate heat itself. Wien's Law states that *the wavelength of radiation is inversely proportional to the absolute temperature of the radiating body*. This means that the greater the heat of a substance the shorter the wavelength will be. Thus the sun emits short-wave radiation while the earth gives off long-wave radiation. It is by means of long-wave radiation from the earth that the atmosphere is heated. Thus temperatures generally decrease vertically with increasing distance from the earth.

Although some of the long-wave radiation is lost to space, a considerable proportion of it is trapped. Water and carbon dioxide in the atmosphere are almost transparent to short-wave radiation but prevent the escape of long-wave radiation. Thus, on a clear night, temperatures fall much lower because there is less water vapour to trap the long-wave radiation, whilst a layer of cloud acts as a blanket confining the loss of heat. The importance can be illustrated by the figures: 47 per cent of incoming radiation passes through the atmosphere to earth, while only 7 per cent of long-wave radiation escapes directly back into space. The trapping of long-wave radiation by water and carbon dioxide is termed **the greenhouse effect**. The importance of this will be examined in more detail in respect of pollution in Chapter 8.

There are other means of heat transfer apart from radiation.

Latent heat

Variations in the effectiveness of heat transfer within the atmosphere can result from the storage or release of **latent heat**. The energy required to convert water from a liquid to a gas is 'stored' in the water vapour as latent heat. This causes a reduction in the temperature of the air. Conversely when water vapour is re-converted to water this latent heat is released so raising the temperature of the air.

Because of this process it can be said that the water vapour content of the atmosphere represents a vast storehouse of heat, in latent form. The transfer of heat in this form takes place both vertically and horizontally. When air containing water vapour rises and reaches **dew point**, the temperature at which vapour condenses, this heat is released. Horizontally, the movement is towards both Equator and poles from the region of maximum evaporation (centred about 25°N and 25°S).

Conduction

Heat transfer also takes place by **conduction**, which is the movement of heat from one substance to another by contact. The earth warms the layer of air in contact with it, and that layer of air transmits heat to the layer above, and so the heat gradually becomes diffused vertically through the air. This process is also responsible for the transference of heat downwards. The surface layer of the earth warms the layer beneath it.

Because air is a poor conductor, only the lowest levels are warmed through conduction, so the process is responsible for only a small proportion of the heating of the atmosphere.

Convection

Unlike latent heat and conduction, the transfer of heat by **convection** can only take place vertically, not horizontally. Warm air rises, and this is termed a **convection current**. As warmed air rises it transfers heat to higher layers. More air is drawn in underneath to replace the ascending air. This is warmed and then rises. The process of convection is important in the formation of many weather phenomena like clouds, storms and winds.

It is important to remember that heat is transferred horizontally over the earth's surface as well as vertically in the atmosphere. Look back to Figure 1.7, which shows that the receipt of radiation is higher in low than in high latitudes. Some of the heat produced by the higher amounts of radiation is transferred horizontally to reduce the imbalance between regions.

The graph in Figure 1.9 shows that there is a surplus of energy in the region between 30°N

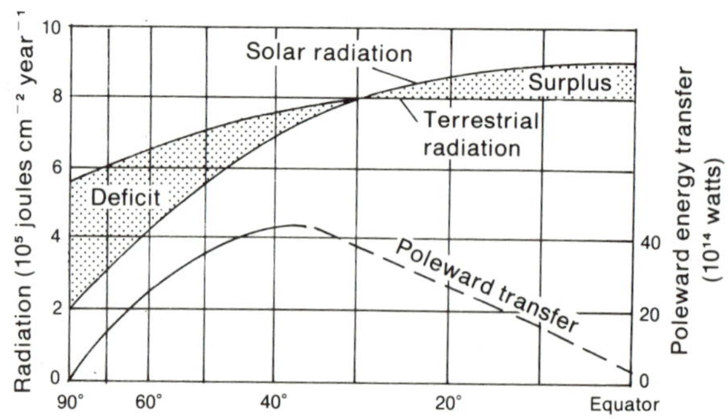

Figure 1.9 The surplus and deficit of the energy budget

and 30°S. Poleward from the 30° latitude there is an energy deficit. To maintain the global energy balance there is an exchange of heat. Equatorial heating initiates this exchange. As the warm air rises, carrying large amounts of latent heat in the form of water vapour, it expands to create high pressure in the upper atmosphere so that at high altitudes air flows south and north outwards from this high pressure zone. As this air meets the cooler air of the middle latitudes, the water vapour condenses and so latent heat is released.

The hydrological cycle

As we have seen, the presence of water influences the energy budget, particularly as part of the process of heat transference through latent heat, and in its effect on the rate of absorption of incoming radiation. Like the energy budget, **the hydrological cycle** is also a subsystem within the climate system, with energy fuelling the system.

The circulation of water from the oceans through the atmosphere back to the oceans —either directly, or indirectly from the land through overland water courses and subterranean passages—is shown in Figure 1.10. The hydrosphere may be considered as a system with four reservoirs between which there are flows. These reservoirs are the oceans of the world, the terrestrial waters (like lakes), polar ice, and water present in the atmosphere. The amount stored varies from time to time both on minor and major scales.

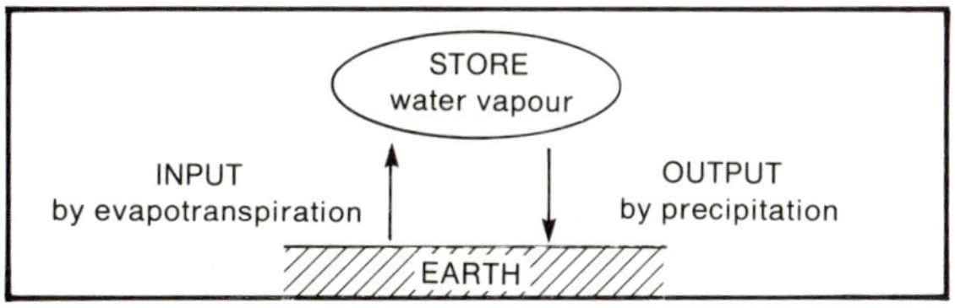

Figure 1.10 The hydrological cycle

The processes by which water flows between the various reservoirs are indicated in Figure 1.11, and are as follows:

Evaporation

Incoming short-wave radiation affects water molecules and if there is sufficient radiation the molecules fly off into the atmosphere as water vapour. These molecules retain some of the energy they have absorbed, and it is this energy which is termed latent heat, described earlier. Evaporation is therefore a process by which solar energy is transformed into latent heat energy, and is also the process which maintains the hydrological cycle.

Transpiration

Transpiration is similar to evaporation and has the same result—the release of water into the atmosphere. Soil water is absorbed by plant roots and is then drawn up through the plants, a process known as **osmosis**. Finally, it is released through the stomata, which are tiny pores in leaves, and the loss is replaced by the continued drawing up of water through the roots. Solar energy plays a part in this because the upward transfer of water is set in motion by the warming of leaf surfaces by the sun. Evaporation of the plant moisture in the leaves triggers off the process of osmosis. Transpiration, because it is initiated by solar radiation, can only take place during the day.

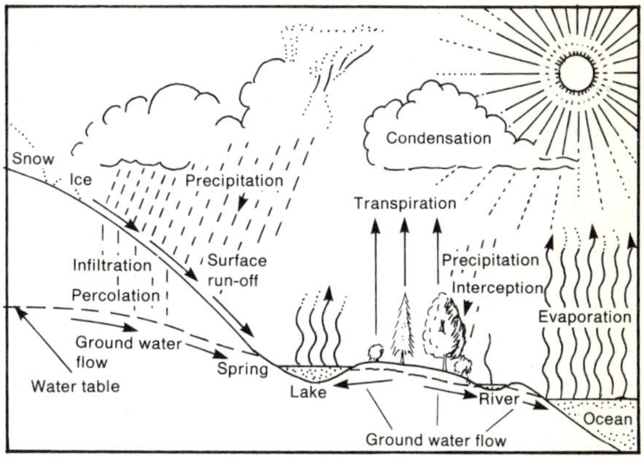

Figure 1.11 Processes of the hydrological cycle

Because it is impossible to measure the amount of water vapour produced by evaporation and by transpiration separately, the term **evapotranspiration** is used to link the two processes. Once water vapour has been released into the atmosphere, it is then stored until such time as the next process in the cycle takes place.

Condensation

Water vapour condenses on a cold surface, as illustrated by early morning dew. Normally air cools because it rises to areas of lower atmospheric pressure: this is termed **adiabatic cooling**. As the air cools, its ability to hold water vapour decreases. Eventually **saturation** occurs, when the air is holding the maximum possible amount of water vapour for its temperature. It is then that condensation takes place, provided **condensation nucleii** are present. These are small particles suspended in the atmosphere, like dust, smoke, sulphur dioxide and salt. Water droplets form around these condensation nucleii and combine with others to form raindrops.

Stability and instability

Air is described as **stable** if it resists vertical uplift and, if forced to rise, tends to return to its former position. When the upward displacement of air results in a tendency for the air to continue to rise, the air is termed **unstable**. We have examined adiabatic cooling, and it is important to recognise that the rate of cooling depends on how stable or unstable the air is. The degree of stability of any column of air is in turn dependent on temperature, since temperature controls the expansion and contraction of the air.

Dry air when forced to rise cools at the rate of 3°C per 300 metres. This is described as the **dry adiabatic lapse rate** (DALR), the lapse rate being the rate of temperature change with height. When saturated air rises, it cools as it ascends at the rate of only 1.5°C per 300 m because of the release of latent heat through condensation. When the lapse rate of a column of air is less than the surrounding lapse rate (which is known as the **environmental lapse rate**), the air is termed stable. When the local lapse rate is higher than that of the surrounding air, the air is unstable.

The degree of stability or instability is important because it affects the processes within the hydrological cycle. Unstable air is more likely to result in condensation and the formation of clouds, which in turn may lead to precipitation.

Clouds

Clouds appear in a variety of forms depending on the precise nature of the processes at work in their formation. They consist of millions of suspended water droplets. Clouds can be classified according to height, appearance or mode of formation.

(a) Height. There are three principal levels of cloud:
high cloud: 6000–12 000 m
medium cloud: 2000–6000 m
low cloud: below 2000 m

(b) Appearance. Shape and properties are the main bases for division, and Luke Howard in 1803 applied a series of Latin terms to them:
cirrus – hair
cumulus – a heap or mass
stratus – a layer
nimbus – rain or storm.
Figure 1.12 illustrates the principal combinations of cloud types.

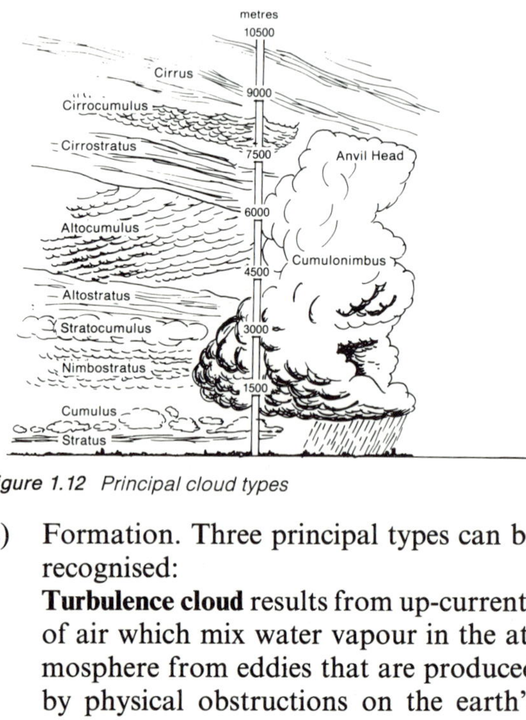

Figure 1.12 Principal cloud types

(c) Formation. Three principal types can be recognised:
Turbulence cloud results from up-currents of air which mix water vapour in the atmosphere from eddies that are produced by physical obstructions on the earth's surface.
Orographic cloud is also a result of relief of the land. Air is forced to rise over mountain barriers and the air is cooled during uplift, with clouds forming if dew point is reached. How often do photographs of mountains appear to have the peak in cloud? Mount Robson in the Rocky Mountains is rarely seen in its entirity for this reason.

Figure 1.13 *Orographic cloud around the peak of the Matterhorn*

Convection cloud – air rising as the result of heating may ascend sufficiently high for dew point to be reached. If the air is very unstable, then great vertical thicknesses of cloud may develop.

Fog may be considered a form of low cloud, and results either from cooling or from evaporation. If cooling takes place through long-wave radiation from the earth, the resultant suspension of water droplets close to the earth's surface is known as **radiation fog**. When warm, moist air crosses a colder surface, **advection fog** can result. Other types of fog include **upslope fog**, when air moves up mountain sides; and **mixing fog**, when warm and cool moist air combine. These are the most common types. Fog formed through evaporation is less common, and may occur over sea areas when warm water evaporates into cold air, and as **frontal fog** when rain along a front evaporates in warm dry air beneath the cloud, but then condenses again in cold air below the front.

Precipitation

Precipitation is a general term which includes rain, drizzle, sleet, and solids like snow and hail. It is the part of the hydrological cycle that returns water to the earth's surface. It occurs when condensation has taken place in the atmosphere and conditions there cannot support the quantity of water droplets or ice crystals in suspension.

Precipitation feeds the three surface reservoirs of the earth: as snow compacting on the polar ice (see D. Ford's text *Ice, Land and Man* in this series for further detail); by falling directly into the oceans, seas, lakes and rivers of the world; and by falling onto the ground and being transported overland as rills, streams and rivers, or percolating beneath the surface to be stored as ground-water.

Precipitation is not evenly distributed throughout the world, and regions of high and low precipitation can be easily distinguished on the map, Figure 1.14.

Study questions

1 List the principal regions of maximum precipitation shown in Figure 1.14. What do they have in common?
2 List the main dry regions of the world. Are there any factors common to their location?
3 Can you discern any patterns of increasing and decreasing precipitation amounts? If you can, describe them briefly and attempt to explain them.

Study activity

Where have all the trees gone?

The forests of the world are an enormous natural resource, and not only in terms of the timber they provide and the tree products that are gathered from them. They also have an important effect both on the global energy budget and on the hydrological cycle. Table 1.2 shows that forests have a low albedo, while in discussion of the hydrological cycle, the importance of transpiration was described.

The Amazon Forest of South America is the largest area of tropical rain forest in the world, covering something like 6.6 million square kilometres. According to satellite photographs, about 100 000 square kilometres of this forest were lost in 1975 alone, representing nearly 4 percent of the total forested area. If the rate of loss were to continue at that level, by the year 2005 the forest will have disap-

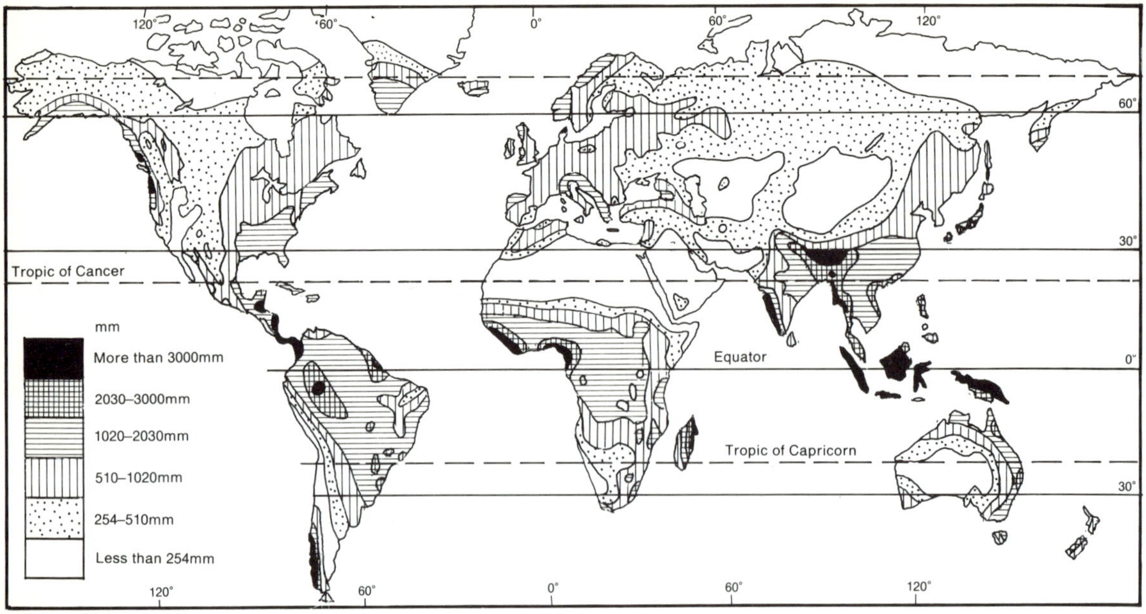

Figure 1.14 Global mean annual precipitation

peared. Even with government controls, 40 percent of the forest can be cleared without breaking any laws.

What happens as the forest disappears? These are some of the possible changes:

1 Tropical forest plays a major role in converting carbon dioxide into oxygen because photosynthesis occurs throughout the year.

What are the effects of significantly reducing the conversion of carbon dioxide into oxygen on the constitution of the atmosphere?

2 The removal of a forest cover results in the degrading of soils through leaching and the increased speed of run-off leads to a higher risk of flooding. The loss of soil fertility aids the production of deserts. Forests generate rainfall far more effectively than other vegetation surfaces because there is a much higher rate of transpiration and because moisture is trapped for longer in the greater vertical extent and increased complexity of the **biome** (a vegetation community of distinctive plant groups). As a result drainage is slower and so there is a greater effective area providing a source of moisture to the atmosphere through evaporation and transpiration.

What are the effects of removing forest cover and thereby reducing the supply of water vapour to the atmosphere?

3 Grassland has a higher albedo than forest, so if the forest is replaced by grassland, the amount of insolation reflected back, i.e. lost to the atmosphere, is higher. Thus less radiation is being absorbed.

As a result of removing the forest cover, what effect will the change in the albedo have on the global energy budget?

4 The results of the disappearance of Amazonia as a tropical rain forest region could be catastrophic in terms of the balance of the climate system.

What kinds of legislation would you suggest that could both protect the forest but also permit the exploitation of its valuable resources by the countries in which it is located?

2

Spatial changes of climate

The study of climate is the legitimate province of the geographer not only because of its effect on the earth's physical environment and people's activities, but also because it exhibits spatial variation. The climate experienced at any one location is unique to that point; at no other location will an identical climate occur.

Climate frequently varies over very short distances, particularly in areas of diversified relief. Valley bottoms in Britain are notably prone to frosts, as heavy cold air sinks during the night and accumulates at the lowest point. A distinctive 'tide-mark' of frost damage along the valley sides may mark the upper level of frost, and the valley slopes above may be 5–6°C warmer than the bottom of the valley. On warm summer days the same valley bottom may be much warmer than the exposed slopes above it, as the air is heated by contact with the ground surface. Sometimes, mysterious, almost circular, flattened areas can be seen in fields where cereals are growing. They are thought to be formed by small whirlwinds, existing for only a few seconds, that can flatten wheat into a circle with a diameter of about 10 metres. We call these features **micro-scale**. You can find examples for yourself by undertaking the project described in the study activity on this page.

Study activity

Hypothesis: *Temperature will vary across a valley cross-section in relation to the height of the land and to the aspect.* The aim of the project is to test this hypothesis, using the following methods:

(a) Choose a valley that is accessible to you.
(b) Take temperature readings across the valley cross-section for a whole day. Select a series of locations from the valley sides and bottom and, for each site, record the air temperature, the time, the location and the weather conditions. Note that you must keep your thermometer out of direct sunlight and take your readings in open spaces. Wait a few minutes at each survey point to ensure that the thermometer is recording a true reading.
(c) When you have completed the cross-section once, leave an interval and then repeat the procedure once or twice more.
(d) Draw a cross-section of the valley from the Ordnance Survey map, marking the points where you took temperature readings.
(e) Using the same scale as the cross-section drawing, plot line graphs to show the temperature you have recorded on each transect.
(f) Examine the graphs in relation to the cross-section. Does the information you have obtained support the hypothesis? Are temperature differences greater at any particular time of day? Does aspect or altitude appear to be more important in determining temperature differences?

Although Britain is quite small, there are large differences of climate to be found between north and south and between east and west. The mild, damp but often windy climate of Cornwall is very different from the drier climate of East Anglia where colder winters and warmer summers produce what is described as more continental conditions. At sea level at Great Yarmouth in Norfolk, the average annual precipitation is 625 mm while at sea level on the coast of Gwynedd in Wales, in the same latitude, it is more than double this amount. On the west coast of Wales, rain-bearing weather systems arrive from the Atlantic without having to cross several hundred kilometres of land, as is the case in East Anglia. Climatic differences of this magnitude can be described as medium or **meso-scale**.

When viewed from a global level, all of Britain has a very similar climate, and it is the large or **macro-scale** differences, for example between the tropics and the temperate areas, that stand out. In a similar way individual atmospheric systems vary in scale: a shower may affect only a few square kilometres and have a lifetime of less than an hour, while a cluster of thunderstorms may affect several hundred kilometres and last for many hours. Larger weather systems like the trade winds or the monsoons of Asia will have typical dimensions of many thousands of kilometres. These scales are subjective in application because many weather systems vary tremendously in extent, but the classification, if not applied too rigidly, is a convenient framework in which to study spatial climatic variations.

Regional classifications of climate

The classification of climates is difficult because climatic parameters are a continuum, without natural divisions, so every division is to some measure arbitrary and subjective. Besides this, there are so many aspects that contribute to the climate of any one location that it is necessary to isolate those that appear to provide the most distinctive features of climate. Some kind of primary simplification is required before commencing classification, because the quantity of data is so vast. Usually, data is simplified by the calculation of mean monthly or annual values, monthly and annual totals. Such averages, while making the work of classification easier, may at the same time obscure or distort significant variations in climate. There are considerable variations in the amount of data available from different parts of the world. There are few weather stations for the ocean areas and some parts of the tropics, in contrast to regions like Western Europe and the United States which have a fairly dense network of stations. Statistical processing of the data will give a clearer picture of the situation in regions with a dense network of stations than in regions with a very sparse covering.

Climatic classifications

Although there are many climatic classifications, there tend to be only two fundamental approaches. One is **genetic**, based on factors that determine or cause different climates, like air circulation. The other is **generic** or empirical, based on climatic observations or on the effects of climate on vegetation or man.

Classification of world climates is necessary because an understanding of the distribution of climate around the world is a first step to understanding how the climate system works. Attempts at qualitative classification of climates began in the nineteenth century when it was recognised that the vegetation types of the world reflected the climate since plants respond to the amount of water available, the hours of sunshine and the temperature. Köppen (1846–1940), a German climatologist, published his version of world climatic regions in 1900. He used vegetation units as the basis of his classification. Vegetation usually changes gradually, but sudden variations may reflect controlling factors like soil type and topography; and Köppen's classification is often criticised for this reason. He divided the climate of the world into five major zones according to temperature. Each of these was then subdivided on the basis of seasonal rainfall and extreme monthly temperature, as can be seen in Table 2.1.

The temperature limits used were arbitrary and were determined from the study of vegetation groupings. Later, Thornthwaite (1899–1963) published his classification in 1948. He tried to devise an improved classific-

Table 2.1 Köppen's algebraic classification of world climates

	Five major temperature groups	Subgroups based on rainfall	Second sub-division
A	Tropical (Average monthly temperature > 18°C. No winter. Annual rainfall exceeds evapotranspiration)	m = short dry season of monsoon type f = no dry season w = winter dry season	
B	Dry (Annual evapotranspiration > precipitation)	S = steppe (38–76 mm year^{-1}) W = desert (< 25 mm year^{-1})	Dry hot (h) = mean annual temperature > 18°C. Dry cold (k) = mean annual temperature < 18°C
C	Warm temperate or mesothermal (One month > 10°C. Coldest month < 18°C and > −3°C)	f = no dry season w = winter dry season s = summer dry season	a = warmest month > 22°C b = warmest month < 22°C c = less than 4 months > 10°C
D	Snow or microthermal (Coldest month < −3°C. Average temperature of warmest month > 10°C)	f = no dry season w = winter dry season	a = warmest month > 22°C b = warmest month < 22°C c = less than 4 months > 10°C d = coldest month < 38°C
E	Ice (Average temperature of warmest month < 10°C)	T = tundra F = permanent frost	

ation which would be useful for agriculture and forestry development in the USA. As the basis of his work he developed a moisture index which he hoped would summarise the availability of moisture in the soil for plant growth. The result is very complex and difficult to use as a world classification.

Recently climatologists have agreed that classifications like those of Köppen and Thornthwaite are misleading, based as they are on mean annual values, since in effect vegetation is probably adjusted more to the extreme aspects of climate like drought and frost frequencies than average aspects.

Study questions

Study Table 2.2 carefully and answer the following questions:

1 How significant do you think are the differences in climatic features between the regions within: (a) the hot climates; (b) the warm temperate climates, and (c) the cool temperate climates?

2 As a result of your analysis for question 1, evaluate the effectiveness of the classification.

3 Examine the column headed *Natural vegetation*. How far do you think that the vegetation types are related to the climatic types?

4 Does there appear to be much relation between the different climates and different ways of life and types of agriculture?

Human classifications

To classify the climates of the world from the human standpoint has been recognised as more useful and relevant to modern requirements. People have long been aware of climatic differences in regard to their physical comfort and have reacted accordingly. A classification of climatic environment based upon human physiological reactions could be of help to medical geographers advising on nutritional

Table 2.2 The major climatic regions of the world

	Location	Features
A	**1 Equatorial** Within 7° of Equator near sea-level. Amazonia, Zaire, Indonesia, Malaysia	Hot and wet all year. Annual range: 1–2°C, diurnal range greater. Mean annual temperature *c*. 27°C. Convection I.T.C.Z.
	2 Tropical From 7° to *c*. 20°N. and S. Includes: (a) coasts with onshore trade winds, e.g. W. Indies, Brazil, E. Africa, E. Australia;	Temperature over 25°C. Rain at all seasons, with summer maximum from easterly waves.
	(b) the continental interiors of tropical S. America, Africa, Australia (called 'Sudan' type).	Hot summers *c*. 28°C or more with convection rain. Winters warm and dry.
	3 Tropical monsoon S. and S.E. Asia, N. Australia	Seasonal wind reversal leading to mild dry season, and hot season, during which maximum rainfall occurs.
	4 Hot desert Between 15–20° and 30°, N. and S. In central and western parts of land areas, e.g. Kalahari in S. Africa, Atacama in Chile, also Sahara, Arabian deserts, Thar, central Australia.	Very hot in summer but with large annual and diurnal ranges. Very dry, rainfall that occurs is sporadic and highly variable. Generally clear skies because of anticyclonic conditions.
B	**1 West margin** 'Mediterranean' type, on west coasts of continents, between 30–40°N. and S. Central Chile, California, S.W. Africa, S.W. Australia, around the Mediterranean.	Mild winters with cyclonic rainfall. Hot summers usually accompanied by drought.
	2 East margin From 23° to about 40°N. and S. called 'Gulf' or 'China' type. Found S.E. USA, Natal, S.E. Japan and China, Uruguay, S.E. Australia.	Mild winters (10–15°C) and hot summers (21–26°C). Rainfall maximum in summer, cold spells occur in winter, especially in S.E. USA and China.
	3 Continental interior Between 30–45° N. and S. Central China; plateau of S. Africa; Murray-Darling plains of Australia, N. Argentina, Paraguay.	Long cold winters (4–8°C) but warm summers (over 25°C). Summer maximum for rainfall which is convectional. More equable in southern hemisphere.
C	**1 West margin** *(Maritime)* W. sides of continents from 40 to 55–60°N. and S. N.W. Europe, S. Chile, British Columbia, Tasmania, S. New Zealand.	Equable temperatures. Precipitation throughout the year. Prevailing westerly winds. Great day-to-day variability.
	2 East margin From 35–40° to 55°N. Eastern USSR and E. Canada. Sometimes called 'Laurentian'.	Cold winters (often below 0°C) and warm summers (*c*. 20°C). Precipitation well distributed through the year.
	3 Continental Interior In N. hemisphere only between 30–40° and 65–70°N. Much of Mid-USSR and China, Canada.	Cold winters (−10 to 15°C) and warm summers (20–25°C). Precipitation light with summer maximum.
D	**Polar** Found in high north and south latitudes. A distinction can be drawn between areas of permanent snow cover (ice-cap) and zone of permafrost (tundra).	Cold throughout the year, although some warm days in summer in Alaska and N. Siberia. Growing season very limited or non-existent. Low annual precipitation.
E	**Mountain** Found in regions of high altitude. Most marked in the fold mountain systems of the world.	Temperature decreases with height. Lowest areas of mountain regions have climates similar to their surroundings; highest areas, similar to polar climates. Precipitation is greatest on the windward slopes of middle height, lower on lee slopes (rainshadow effect) and above 2000 m. Local winds are important.

c. = circa, indicates an approximate value.

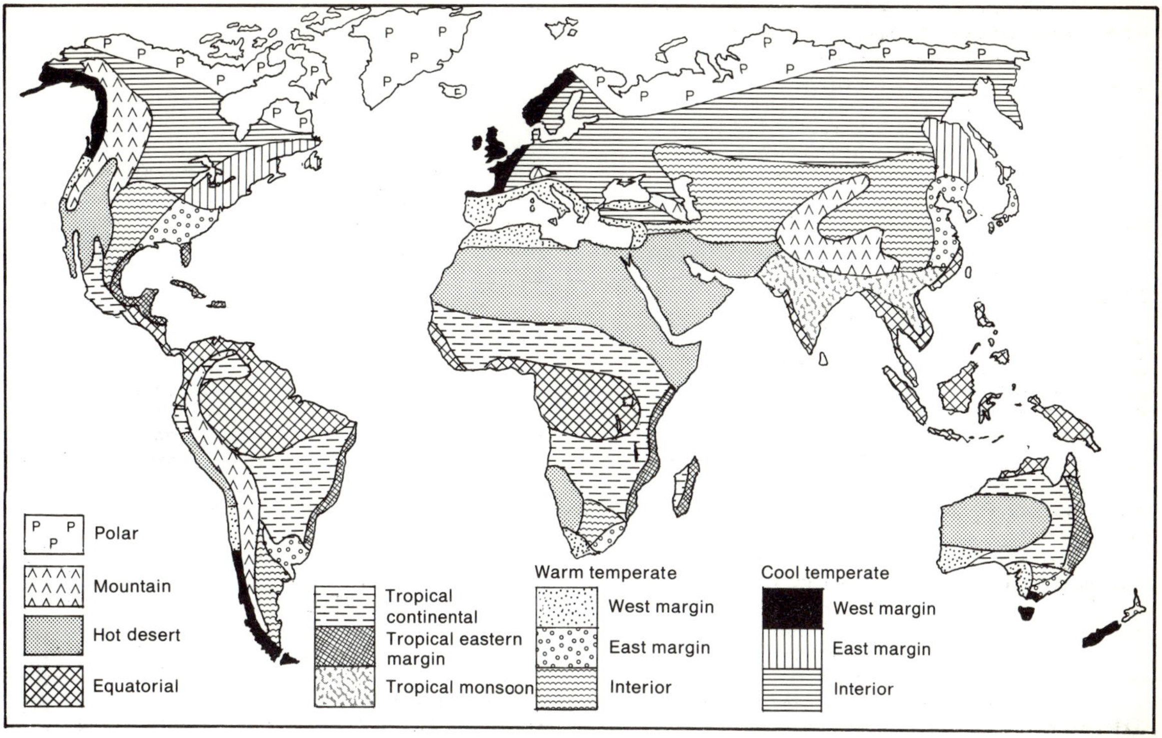

Figure 2.1 Global climatic regions

needs, to holidaymakers, to military strategists planning clothing requirements for a fighting force, and to many others. It is perhaps surprising that little work has been done on classifying the climate with which we have to live.

A classification based on how comfortable we human beings feel would be desirable. Unfortunately such a classification presents many difficulties. What constitutes a good climate? The 'ideal' climate in which to spend a holiday is not necessarily the ideal climate in which to work. We shall now examine some of the criteria on which a human classification might be based.

Physiological temperature

Physiological temperature is the temperature experienced by a living organism and depends both on air temperature and the rate of heat loss from the organism, which is affected by evaporation. The efficiency and speed of evaporation is controlled by air humidity, wind speed and exposure to sunshine. Indices of physiological temperature are usually based on air temperature and humidity, but exclude wind and sunshine. An **effective temperature index** (ETI), also known as the discomfort or the temperature humidity index, can be produced from the indices. It can be used to determine the range of comfort zones, dividing the world into regions of varying degrees of comfort or discomfort to man.

Figures 2.2 and 2.3 show effective temperature in Africa in January and in July. An ETI value of 18.9°C or below indicates an uncomfortable condition arising from cold stress and a value of 25.6°C or above indicates discomfort through heat stress.

Study activity

With the aid of the maps (Figures 2.2 and 2.3):

1 Locate those regions of Africa that could be described as 'comfortable' in January.

2 Do the same for July.

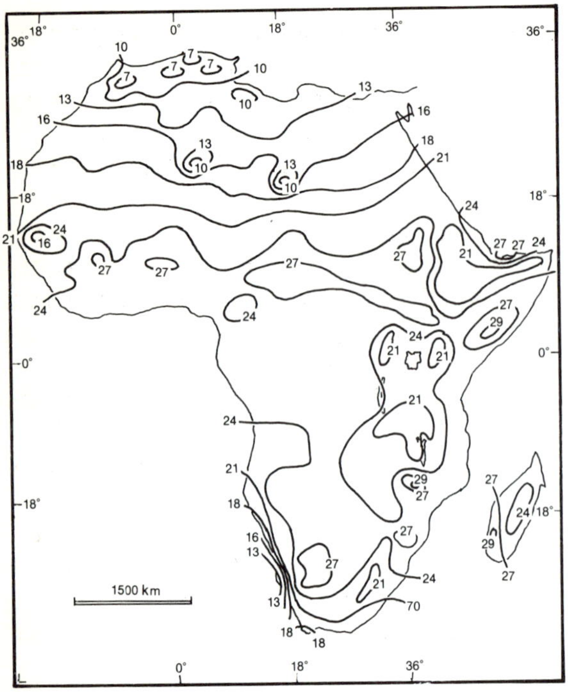

Figure 2.2 Effective temperature in Africa in January

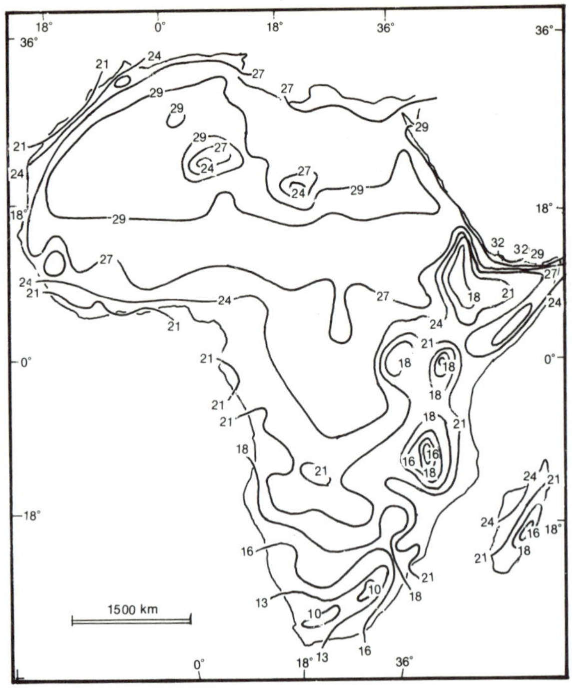

Figure 2.3 Effective temperature in Africa in July

3 From this information, divide Africa into the following regions:
(a) comfortable in January and July;
(b) comfortable in January but uncomfortable in July through cold stress;
(c) comfortable in January but uncomfortable in July through heat stress;
(d) comfortable in July but uncomfortable in January through cold stress;
(e) comfortable in July but uncomfortable in January through heat stress;
(f) uncomfortable in January and July.

4 Attempt to produce a map of Africa to show the location of these regions.

5 With the aid of the map in Figure 2.4, describe any relationships between the distribution of population and degrees of human comfort.

Figure 2.4 Population density of Africa

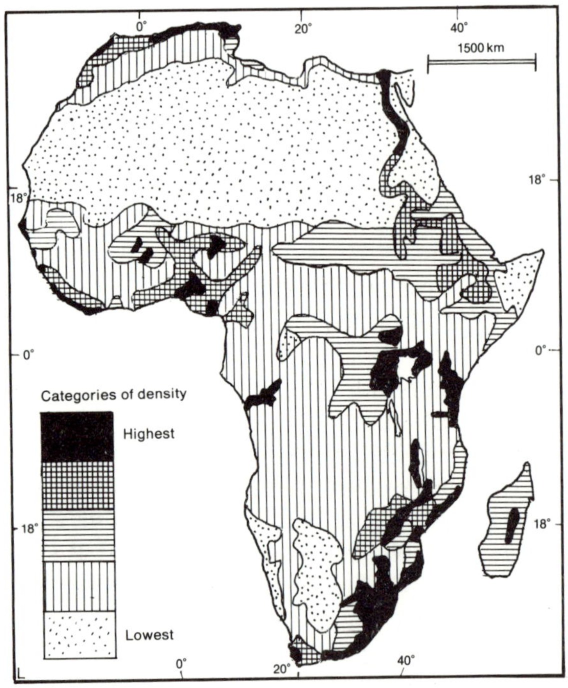

Clothing zones

Because climate influences the way we dress, clothing varies not only from one culture to another but from one climatic zone to another.

It is therefore possible to use clothing types to classify climatic regions. Seven zones based on clothing requirements have been recognised:

1 The minimal clothing zone—the humid tropics with mean temperatures between 20° and 30°C.

2 Long, flowing robe zone—the hot deserts with high daytime temperatures but lower night temperatures.

3 The one-layer clothing zone—the subtropical belt with mean winter temperatures between 10° and 20°C.

4 The two-layer clothing zone—the temperate regions with cool winters of between 0° and 10°C mean temperatures.

5 The three-layer clothing zone—the cold winter belt of the temperate regions with mean winter temperatures between −10° and 0°C.

6 The four-layer clothing zone—the temperate regions with severe cold winters, between −20° and −10°C.

7 The Arctic zone—the region where a comfortable body-heat balance cannot be maintained by clothing alone.

Housing zones

Buildings offer protection from the weather, and houses should have an artificial climate that makes living in them a pleasant and comfortable experience. To achieve this, traditional building styles have evolved which are well adapted to the environment. Unfortunately modern building styles, like high-rise tower blocks, have been exported from temperate to tropical areas with little regard to their suitability. It often requires a great deal of money spent on air conditioning or double glazing to make such buildings habitable. Thus while the skylines of Lagos, Rio de Janeiro or Kuwait bear a considerable resemblance to those of Paris, London or Chicago, it is often the case that living or using imported-style buildings in the tropics can be very uncomfortable. By considering native housing styles, which are often well adapted to the environment, we can recognise seven main zones:

1 Hot humid zone—the houses are often raised and of one-room depth. By being raised, a measure of protection against animals and insects is achieved and ventilation is improved.

2 Hot dry zone—houses are often constructed of dried mud bricks, partly because of the unavailability of other materials but also because they are good insulators. Windows are small and roofs flat, and buildings may be painted white to reflect back radiation. Protection from daytime heat is a prime motive, together with retaining heat on cold nights.

3 Hot savanna zone—because this zone exhibits features of both the hot humid and dry zones, dwellings are generally constructed of mud and grass, often under the shade of trees.

4 Mediterranean zone—stone and wood are widely used in construction and the plan of the dwellings often includes an open courtyard. Stone acts as an insulation from radiation, particularly in summer.

5 Subtropical desert zone—temperatures are very high and vegetation very sparse. The traditional way of life is nomadic, using tents made of animal skins.

6 Cool zone—because of the need to conserve heat, dwellings are generally small, and good insulation is necessary, often provided in the past by thatched roofs.

7 Cold zone—in forested areas, wooden huts with steep roofs to prevent too much accumulation of snow were common. In the extreme north where temperatures are very low, the Inuit (formerly Eskimos) developed the igloo style of dwelling, using snow blocks.

The different building styles are illustrated in Figure 2.6.

Study questions

1 What are the main disadvantages of employing classifications like those of clothing and housing?

2 Are there any other aspects of human life that can be used to produce a climatic classification? If so, suggest a method of classifying them according to climate.

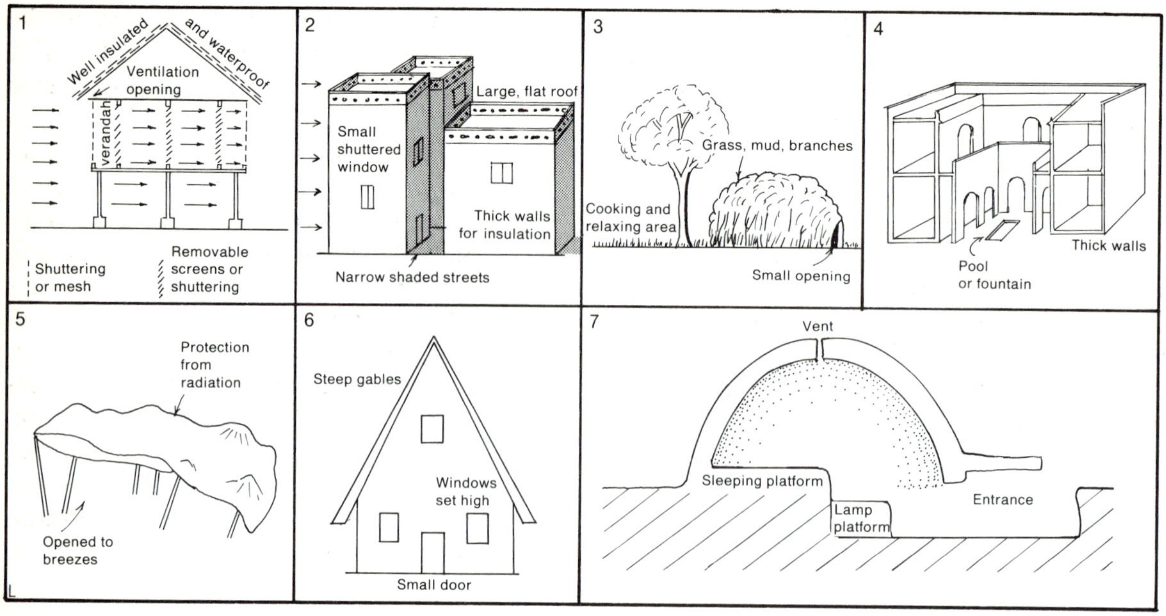

Figure 2.5 Native house types in various climatic zones

Wish you were where?

In the age of jet travel more people than ever before are travelling around the globe on business and for pleasure. In order to enjoy the trip and pack the right clothes the traveller needs to know what the climate will be like at his destination. In the middle of a British winter the thought of the heat of Singapore may be very tempting, but would it be too humid and uncomfortable? Perhaps you have friends who have spent their summer holidays at a Mediterranean resort. What are places like Benidorm or Corfu like in the winter months when the prices are cheaper? The holiday brochures show sunny skies and swimming pools. In this section we look at the climate at a 'long-haul' destination and at the more popular southern European resorts to answer some of these questions.

Singapore—an Equatorial climate

The climate statistics for this Far Eastern city located at 1°N. are given in Table 2.3.

Characteristics

Note how the climate is almost completely lacking in seasonal variation. This is mainly because the sun is never at more than a slight angle from the veritical. Therefore the amount of insolation received here by the earth's surface changes little throughout the year. Temperatures are always high, averaging between 25° and 28°C, and the annual range is very slight (see Figures 1.7 and 1.8 illustrating radiation and temperature). However, diurnal ranges are more pronounced and provide the main thermal variations, so that it has been commented that 'the nights are the winter of the tropics'.

Table 2.3 Climate statistics, Singapore

	J	F	M	A	M	J	J	A	S	O	N	D
Average maximum temperature	30	31	31	31	31	32	31	31	31	31	31	31
Average minimum temperature	23	23	24	24	24	24	24	24	24	23	23	23
Precipitation (mm)	252	173	193	188	173	173	170	196	178	208	254	257

Rainfall is heavy throughout the year and this climate zone is the wettest in the world. Most of the rain is convectional in origin so that at many inland stations there is an afternoon maximum frequency of rainfall because the greatest convection activity occurs at this time. Rainfall is characteristically of high intensity because the warm air is able to hold a great deal of moisture which is released when the air is forced to rise. At any one time the precipitation is often highly localised and falls in thunderstorms. The frequency of thunder is very high, and may occur on between 100 and 150 days per year.

The combination of warmth and wetness leads to high relative humidities which make the climate oppressive in which to live and work. 'Monotony' and 'uniformity' are words that are frequently employed to describe this climate. The cooler nights often fail to bring much relief after the day's heat because the continuance of the high humidity maintains the high physiological temperature.

The controlling weather system

In Equatorial areas pressure is generally low, with convergence taking place over a wide zone, referred to as the **inter-tropical convergence zone** (ITCZ). This zone has an average width of between 5 and 10 degrees of latitude, and contains within it many minor waves and vortices which undoubtedly have an effect on the distribution of convection activity.

Figure 2.6 Mean annual rainfall over the tropical Pacific

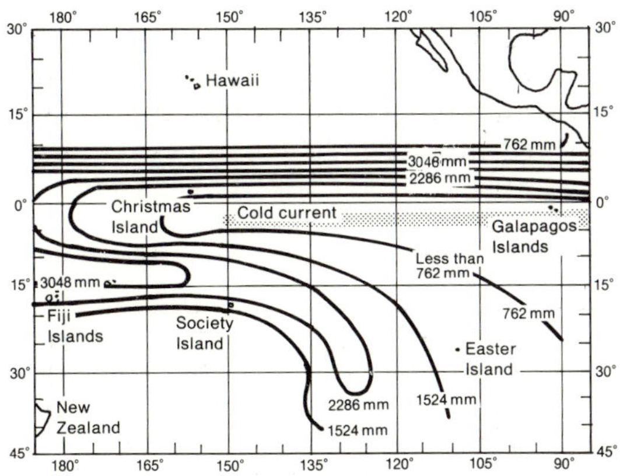

There are variations in the climate of the Equatorial areas as a result of the ITCZ. For example, in the central and eastern Pacific Ocean it is found centred around 5°N, and the mean annual rainfall map of the area, (Figure 2.6), shows that a belt of high precipitation is found here, while immediately to the south, almost along the Equator, rainfall is very much lower and there is a marked dry zone. The position of the latter corresponds to the position of a cold ocean current which has the effect of reducing the convection activity in the atmosphere above it. When, from time to time, the cold current disappears, rainfall becomes much heavier. Figure 2.7 shows clearly how the two are linked, by comparing sea temperature anomalies with the annual rainfall in the region.

Figure 2.7 Tropical rainfall over a 21-year period

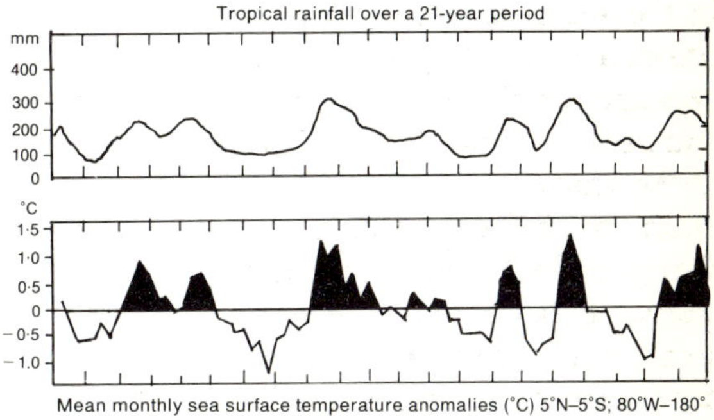

Climate, man and his environment

The nature of the climate influences many features of the biome, including the soils, vegetation and the natural resources that man exploits.

Soils In humid tropical climates, chemical weathering within the soil is more intense than in temperate latitudes. This often leads to the formation of deep profiles, and the *relative* loss of silica results in soils containing quite high proportions of iron and aluminium oxides.

Leaching occurs as a result of heavy rainfall and contributes to the moderately acid nature of the soils. Sometimes the iron compounds within the soil become concentrated to form a horizon termed a laterite, composed of hard concretions of iron.

The name given to most soils found in humid tropical regions is **laterites**. While known as tropical red earths, not all laterites are red, although most have a reddish content. The plant nutrients are contained in the humus, which is constantly renewed by the breakdown of plant material. However, if the forest cover is removed, so that the supply of organic matter is cut off, the humus content rapidly declines and results in a loss of soil fertility.

Vegetation The common vegetation type of the Equatorial regions is tropical evergreen rain forest. Because of the high rainfall and high temperature, the rates of biological activity are rapid and growth takes place throughout the year. Consequently the evergreen rain forest is the most luxuriant of all vegetation types.

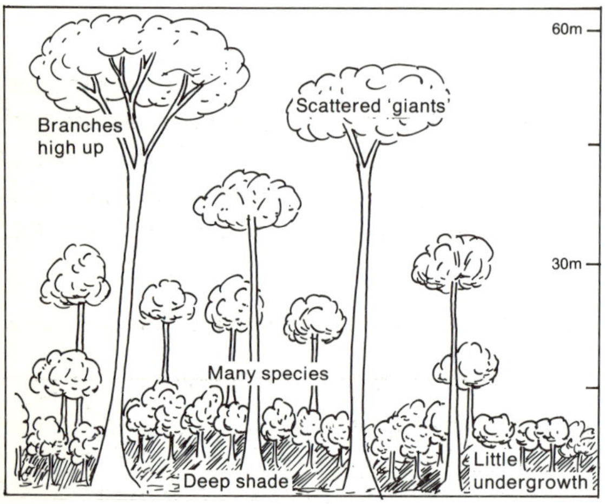

Figure 2.8 *Structure of tropical rain forest*

The rain forest has a recognisable structure (illustrated in Figure 2.8), with about five layers of which the three highest are formed of trees. Below the isolated, very high trees, the second layer is composed of trees forming a continuous canopy with long slender trunks and rounded crowns. Beneath this layer the trees have narrow crowns and the layer is discontinuous, dependent on the availability of light penetrating the canopy above. The field layer consists of herbs, including ferns, but undergrowth tends to be sparse owing to the deep shade. The fifth layer is formed of climbing plants of different varieties, for example lianes, which are woody climbers that are entwined around the tree trunks, often reaching as high as the canopy. Other species include epiphytes which grow attached to the branches of trees, and parasites, that feed off the trees themselves.

Figure 2.9 *Tropical rain forest in the Seychelles*

Exploitation by man As a natural resource, tropical climates are most important in terms of agriculture. About 40 percent of the world's population live in the tropics and all their food and most of their other necessities of life are supplied by agriculture. Tropical agriculture is also important in international trade because many products are supplied that cannot easily be produced in other regions, e.g. coffee, tea.

From the tropical rain forests come many forest products including hardwood timber, and traditionally fruits from the trees that were collected. However, this has since been rationalised so that tree products are frequently produced on plantations. Rubber, from the latex of the hevea tree, palm oil from the West African oil palm, coconuts, cocoa from the cacao tree, and bananas are produced from the widespread cultivation of these trees.

Table 2.4 Climate statistics, Palma and Rhodes

Palma	J	F	M	A	M	J	J	A	S	O	N	D
Average maximum temperature	14	15	17	19	22	26	29	29	27	23	18	15
Average minimum temperature	6	6	8	10	13	17	20	20	18	14	10	8
Precipitation (mm)	39	34	51	32	29	17	3	25	58	77	47	40
Rhodes	J	F	M	A	M	J	J	A	S	O	N	D
Average maximum temperature	16	16	17	20	24	30	32	32	30	25	20	17
Average minimum temperature	8	8	9	11	15	19	21	22	20	15	11	9
Precipitation (mm)	237	123	110	25	26	0	0	0	4	96	152	199

Mediterranean Europe—Britain's favourite escape route to the sun

The average temperature and rainfall values for the two popular Mediterranean resorts, Palma in the Balearic Island of Mallorca, and Rhodes in Greece, are given in Table 2.4.

You can see that hot, dry summers and mild, wet winters are features of both places. The nearly cloudless skies of summer result in a very high total of sunshine hours. Winter is very different—Rhodes has on average more rain than London in all months from November to February. In summer the weather is settled and reliable, with only the occasional thunderstorm to interrupt the succession of warm sunny days. In winter there is more day-to-day variability at both places, but despite the quite heavy rainfall, temperature and sunshine levels are sufficiently high to tempt out-of-season holiday-makers.

The controlling weather systems

Figure 2.10 shows that in summer pressure is quite high over the Mediterranean but in winter the pressure is generally low (Figure 2.11). In winter, depressions form over the Mediterranean Sea and bring unsettled conditions, but although there is a copious rainfall, the intensity of the rain is often greater than we experience in Britain, and mild sunny days occur between the wet spells. Another feature of the climate in the Mediterranean basin is the abrupt change from the settled and dry conditions of summer to the changeable weather of winter.

Climate, man and his environment

Soils The principal soils found in Mediterranean regions are brown Mediterranean soil, terra-rossa and rendzina. Originally most of the vegetation of these regions was probably evergreen forest of both broad-leaved and

Figure 2.10 The general pressure pattern over the Mediterranean in summer

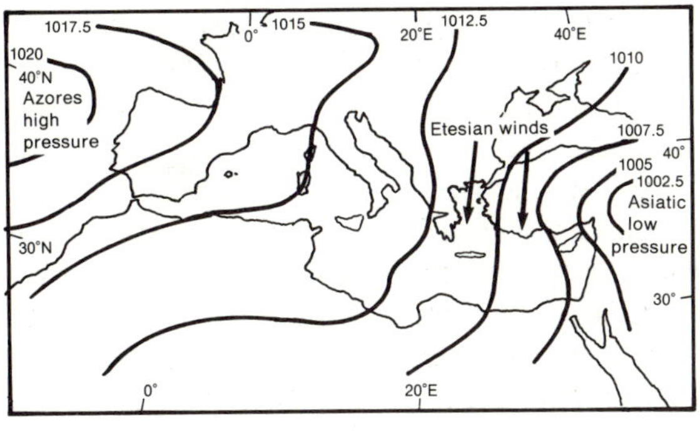

Figure 2.11 The general pressure pattern over the Mediterranean in winter

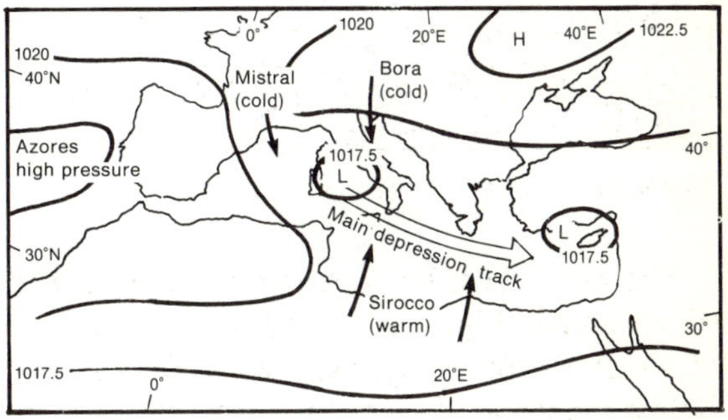

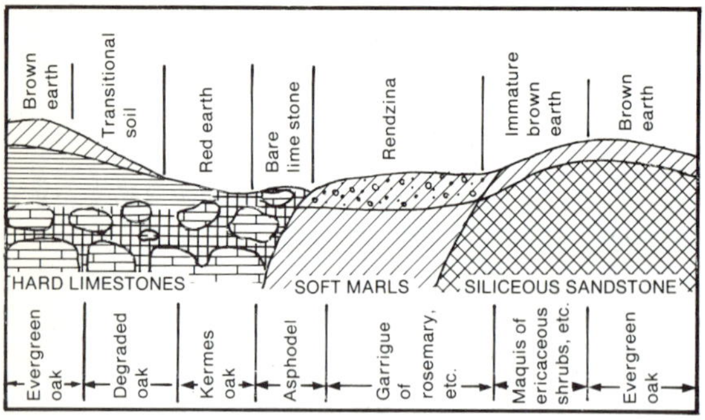

Figure 2.12 Mediterranean associations of soil types and of plant cover

coniferous trees. Much of this forest has been removed by human activity, as a result of which soil erosion has occurred and there have been changes in the nature of the soils.

With the replacement of forest by scrub vegetation the humus content gradually diminishes, and the soils become more acidic. The soils are influenced not only by their vegetation cover but by the parent rock: usually sandstones and shales, or limestones and calcareous marls. On the sandstones brown Mediterranean soils have developed, while weathered limestone has led to the formation of the distinctive terra-rossa, a deep red clay. Rendzinas develop mostly on the marls where there is a sparse vegetation cover. Figure 2.12 illustrates the typical associations of rock, soil and vegetation found in these regions.

Vegetation Mention has already been made of the forests that were thought to have covered much of the Mediterranean regions. They still exist although over limited areas: pines and oaks are the main species, the latter including the cork oak which is economically valuable.

The forests have been replaced by scrub, of which there are two main types: maquis and garrigue. **Maquis**, which occurs on non-calcareous rocks, is quite dense and includes many species of small trees, bushes and plants of which the olive is possibly the best known. **Garrigue** is a more sparse scrub, found on the limestone areas, with bare ground between short plants.

Although the rock type is important as a factor in influencing the vegetation cover, the nature of the climate is of particular significance. Plenty of moisture is available to plants in winter but during the summer the upper horizons of the soil dry out because of the low rainfall, low relative humidity and high evaporation rates. In order to survive, plants have to be drought-resistant. This is why many of the trees are evergreen, and the variety of plant species limited. Limestone areas support garrigue rather than maquis because the limestone dries out more quickly and so there is less moisture available to plants during the drought period.

Figure 2.13 Maquis vegetation in Southern France

Figure 2.14 Garrigue vegetation in Southern France

Exploitation by man Mild winters with adequate moisture, and high summer temperatures, have helped Mediterranean regions to be very productive agriculturally from early times, especially where irrigation schemes have been introduced. The Fertile Crescent, extending from northern Egypt through the Middle East, gave birth to sedentary agriculture, and vast irrigation schemes were in operation during the time of the Babylonian Empire nearly 3500 years ago. It was in this region that cereals like wheat and barley were first domesticated and cultivated.

Today, Mediterranean agriculture still includes the cultivation of grain, but many areas are most important for the growth of vegetables and the products of fruit trees, notably the citrus fruit trees. Vines are grown extensively on terraced hillsides and in valleys, producing currants, raisins and sultanas as well as wine. In those areas unsuited to cultivation because of altitude, drought or paucity of soil, the olive tree nevertheless thrives, and olive oil is a major product of the region. Also in these poorer regions sheep, goats and pigs are grazed.

Although agriculture has been and remains an important aspect of utilisation of the Mediterranean climate, this same climate has become an economic resource in a quite different field of human activity. Whilst hot, dry summers are not often pleasant to work in, particularly for those engaged in outdoor occupations, they provide an ideal environment for relaxation, particularly if water is present, either the sea or in swimming pools. Today the Mediterranean region of Europe is the biggest centre of tourism in the world. In barren coastal areas where formerly only olive trees would grow, tall concrete hotels and apartments mushroom and new infrastructures develop. Tourism is not only a source of employment to those engaged in constructing the hotels and in the associated service industries like hotel keeping, running shops, cafés, bars and nightclubs, but it stimulates other areas of the economy, such as providing an increased demand for agricultural products and a market for souvenirs and clothing.

Study questions

1 Study the rainfall statistics for Palma and Rhodes. How long does the drought last in each resort? If you had to take a holiday in November which resort would be likely to offer the better holiday weather?

2 Collect some holiday brochures for both winter and summer holidays in the Mediterranean region. How important is the climatic element in the advertising and description of the holiday locations? Why do you think that in winter brochures temperature and sunshine data is often given, but rainfall figures only very infrequently?

3 Using the climatic data previously given (Tables 2.3 and 2.4), calculate the annual average range of temperature at Singapore, Palma and Rhodes. Add up the total rainfall at each location. Why does Singapore have more rainfall than the Mediterranean resorts? Which climate do you think would be more comfortable to live and work in? Why?

Conclusion

Whether any classification of climate is useful depends on who is going to make use of it. Classifications are necessary to understand and clarify a complex topic, and are a useful aid to study, simplifying and generalising the complicated patterns of climate that exist on the earth's surface. The earth's climatic zones are among the most important characteristics of the planet, having a profound effect on landscape, vegetation, animal life and human culture. Climatic differences are faithfully reflected in soil and vegetation variations, and in one sense the plant can be viewed as a meteorological instrument interpreting the various factors of climate. In this chapter we have seen that different classifications are required for different purposes, that no single scheme is satisfactory for all purposes, and that, as yet, no method is infallible for even one purpose.

3

Temporal changes of climate

A changing climate

In early medieval times Britain produced large quantities of wine. This would not have been possible if the majority of summers had not been sufficiently warm and dry to allow the grapes to mature. Later, in the sixteenth and seventeenth centuries, we know that the River Thames often froze over and great fairs were held on the ice. These two examples illustrate the fact that even within historical times the climate of Britain has varied from what it is today. Climatic changes on many different time-scales, from the decade to the hundreds of millenia, can be seen in Figure 3.1.

This chapter is devoted to studying the evidence of changes in climate over time. The time-scale being considered is not in terms of hours, days or months but over much longer periods of time.

Because the atmosphere is never static, its characteristics vary both from place to place and over periods of time. Changes may be the result of feedback processes within the climate system or they may be induced by factors external to the system. When examining climatic change, short-term variations in weather are not involved. Instead, the composite of weather and weather variability that together produces a recognisable climate is subordinated to apparent fluctuations in the overall climate. Fluctuations in a climate that tend to follow a trend are termed **climatic trends**, while if these fluctuations appear to follow a cycle, they are termed **climatic cycles**. It is only when the climatic fluctuations are sufficient over a long period of time to produce a climate of a differing nature from the former climate that the term **climatic change** is employed.

Figure 3.2 shows the different kinds of climatic variability that can be recognised. At the top of the diagram, line A shows a series of smooth curves varying in a regular manner with distinct maxima and minima. This kind of variability is known as **periodic variation**. Line B shows a marked **short-term fluctuation**, while line C shows how a downward (or upward) trend would appear. Finally, line D shows increasing variations as time passes, indicating increasing variability.

The examination of climatic change requires careful assessment of the length of time involved in the change. Three main classes can be recognised. The **secular** time-scale involves changes that have occurred within the past 500–1000 years. **Historical** changes have a longer time-span, spread over the last 2000 years or so. **Geological** scale refers to changes that have been occurring from prior to recorded history. Changes that occur over a smaller time-span than any in these scales can be referred to as within the **short-term scale**.

This does not mean that short-term changes are insignificant. Even a period of change of between one and three years can have very pronounced effects on human life. Changes in climate very often affect food supply, because changes in weather will affect harvests, often adversely—as where there is a shortage of rainfall or lower than normal temperatures. For example, weather fluctuations in the USSR in the 1960s led to a series of poor grain harvests, which affected not only the Soviet Union but

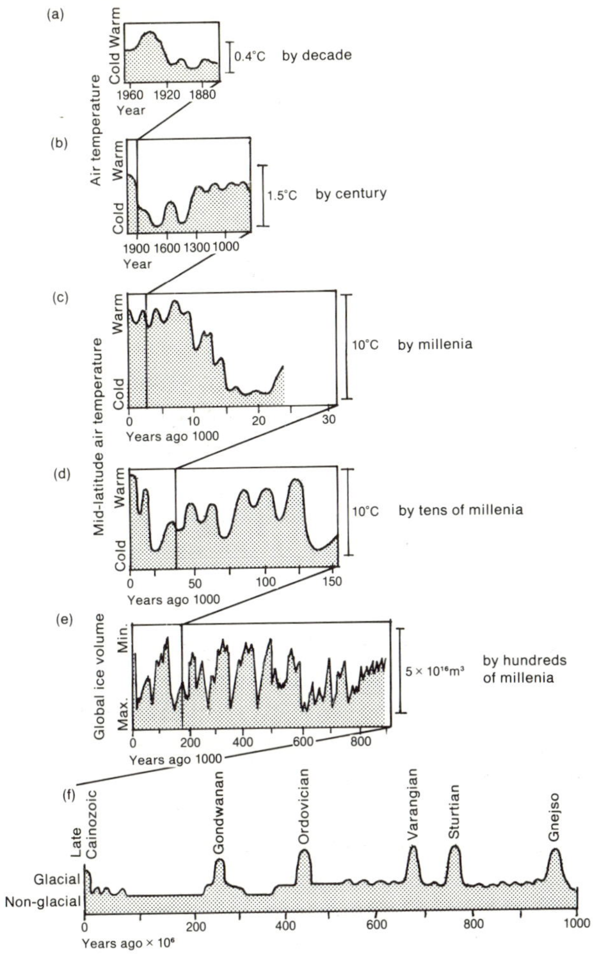

(a) Changes in the five-year average surface temperatures over the region 0–80°N.
(b) Winter severity index for eastern Europe.
(c) Generalized northern hemisphere air-temperature trends, based on fluctuations in alpine glaciers, changes in tree-lines, marginal fluctuations in continental glaciers, and shifts in vegetation patterns recorded in pollen spectra.
(d) Generalized northern hemisphere air temperature trends, based on mid-latitude sea-surface temperature, pollen records, and on worldwide sea-level records.
(e) Fluctuations in global ice-volume recorded by changes in isotopic composition of fossil plankton in a deep-sea core.
(f) The spacing of ice ages in geological time.

Figure 3.1 The different scales of climatic change

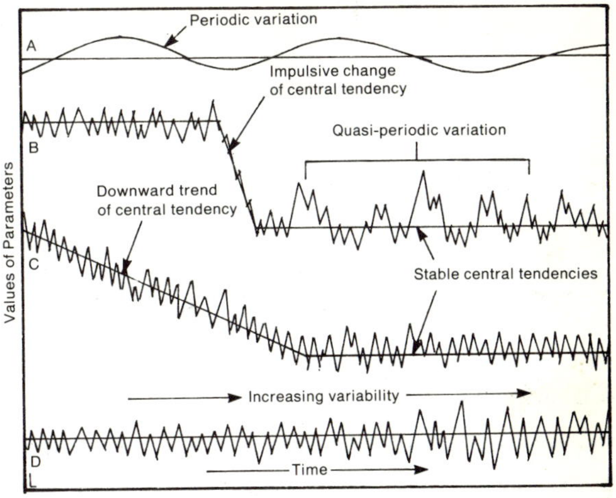

Figure 3.2 Recognisable types of climatic variability

also the rest of the world because grain became more expensive through scarcity. Even worse has been the situation in north-east Africa, where the Sahel drought has now lasted since the late 1960s. Famine, starvation and loss of life, together with political unrest have resulted and the effects on the countries which include part of the Sahel within their boundaries have been disastrous.

In assessing the evidence, two time-scales will be employed. Figure 3.3 gives the scale of climatic changes which can be simplified into **historical** from 2000 years ago until the present, and **geological** for the period preceding historical time.

Figure 3.3 Scale of climatic change

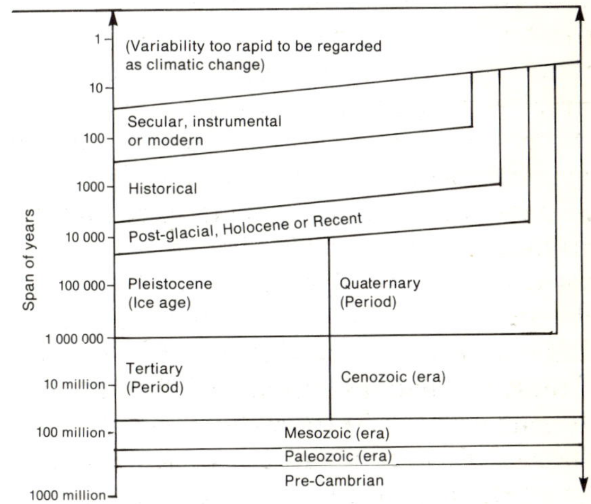

The evidence for a changing climate

Geological time

Evidence that climate in the past has been very different from what it is today in many parts of the world comes in many forms. Different types of evidence need to be assessed for different scales of climatic change. The climate of the geological past can only be pieced together by studying indirect, or proxy, evidence and we shall now examine the main sources of such evidence.

Ice cores

In the highest central parts of Greenland and Antarctica there has been almost no ice melt. Snow has continued to accumulate and be pressed down into ice. Thus a bore can be made through the ice and the sample can include the accumulations of thousands of years. Under a microscope the annual layers can be identified, and net accumulation calculated. Measurement of the acid aerosols and of oxygen isotopes helps to confirm variations. Thus it is possible to determine changes in the atmosphere of sulphuric acid amounts, and to estimate past air temperatures.

Isotope analysis

Isotope analysis is finding increasing application. Isotopes of oxygen^{16} and oxygen^{18} can be measured and their quantities calculated. The former is lighter and is therefore lost more quickly through evaporation, so in high latitudes during winter, and during past colder climates, there are more negative values for oxygen^{18}. From this, fluctuations in temperature during the past can be identified.

Geology

Past climates can also be reconstructed from the study of geological and geomorphological features. Within layers of rocks and sediments, varves, evaporites and weathering processes are all indicators. Varves are distinctive banded layers of silt and sand deposited in lakes near ice-sheet margins. Coarser, lighter coloured material is deposited in summer and finer, darker material in winter. The number of years involved in the formation of a varve can be estimated from the deposits, while their thickness indicates variations in length of winter and summer conditions.

Salt deposits or evaporites are formed in dry, warm conditions when evaporation exceeds precipitation, so their presence is an indication of these climatic conditions. They are particularly important as indicators when they are found in bands of deposits in a climate that is now cool or humid. Laterites are also formed only in warm humid conditions and so provide another source of evidence.

Glacial features, including glacial deposits, moraines and eskers, show the location of former glacial activity and its extent. Surviving land forms like sand dunes give indications of aridity and also of prevailing wind direction. River terraces may also be produced by changes in climate. However, all these morphological features, although easily recognisable, do not provide sufficient evidence on their own of climatic change. A composite of factors is responsible for their formation, of which the climate is only one. They are therefore most important as supporting evidence, and no feature on its own must be taken as sufficient proof of a change in climatic conditions. (You might care to pursue this topic further in D. Ford's text *Ice, Land and Man* in this series.)

Biological phenomena

Plant growth, flowering and pollen productivity are all strongly influenced by climate, and so evidence from the past about changes in vegetation is an important indicator of climatic change. However, like morphological features, there are causes besides climate for changes in plant species and quantity, and so interpretation of biological evidence is necessary. There are three main methods in frequent use in the study of past climates: **palynology** (the technique of pollen analysis), **dendrochronology** (the analysis of tree rings) and **carbon-14** dating.

Palynology Pollen grains are very tiny but nevertheless distinctive. They are carried by the wind, fall to the ground, and become encapsulated in sediments. Because of their lack of oxygen, peatbogs preserve pollen for a long time. A core of peatbog can be extracted, up to several thousand metres long, representing possibly several thousand years of history. The pollen content of each distinctive horizon is analysed and with the assistance of pollen studies for different parts of the world today, a reasonable picture can be obtained of the vegetation associations existing in that place at any particular time in history. From this the relationship between the vegetation and climatic factors for that region can be assessed, so that some tentative conclusions can be reached about the nature of the climate producing that particular vegetation pattern. In recent years, mathematical relationships between vegetational elements and climatic factors have been developed, called **transfer functions**, which aid an increase in the precision of the study. Figure 3.4 shows an example of a pollen diagram from Dundalk, Ireland. Notice how agricultural pollen increases in relative abundance after about 1200 BC as farming increases.

There are two main disadvantages to pollen analysis. Firstly, different pollens have different rates of decay, and so it is possible that they are preserved selectively, like fossils, and may not be representative of the vegetation of any given period of history. Secondly, pollens can be derived from an earlier period and therefore are not indicating the climatic conditions of the time under study.

Dendrochronology The pattern of tree rings helps to establish variations in climate. A tree growing in a benign climate, with adequate water supply and favourable temperatures, does not have much variation in the nature of its annual ring growth. However, if there are considerable variations in temperature and water supply from year to year, these differences are reflected in the width of the tree rings, with narrower rings in drier conditions. It is insufficient to study one tree but if a number of samples are taken from different trees in an area, cross-dating can be attempted. From this a pattern can be established, known as the **mean** or **master chronology**.

The dating of trees is not always straightforward for a number of reasons. There may be

Figure 3.4 A generalised pollen diagram from Red Bog, west of Dundalk, Ireland

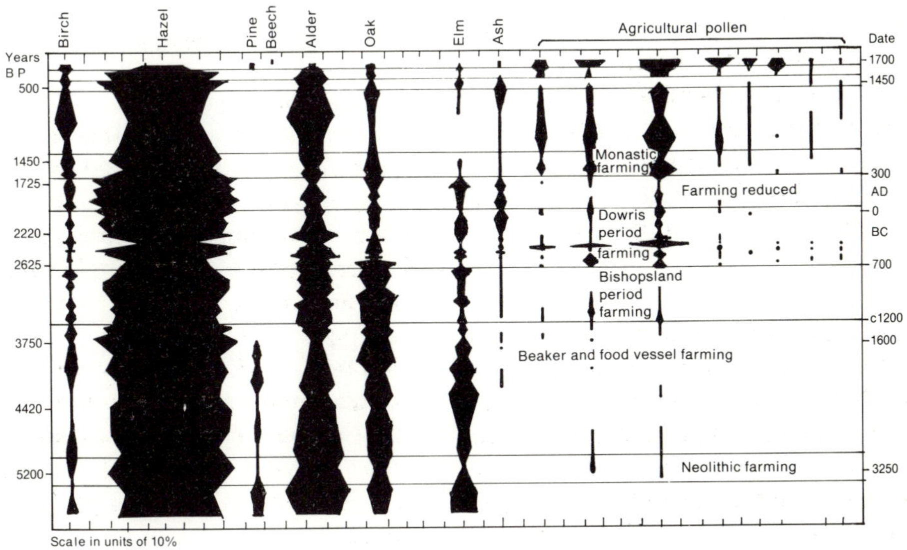

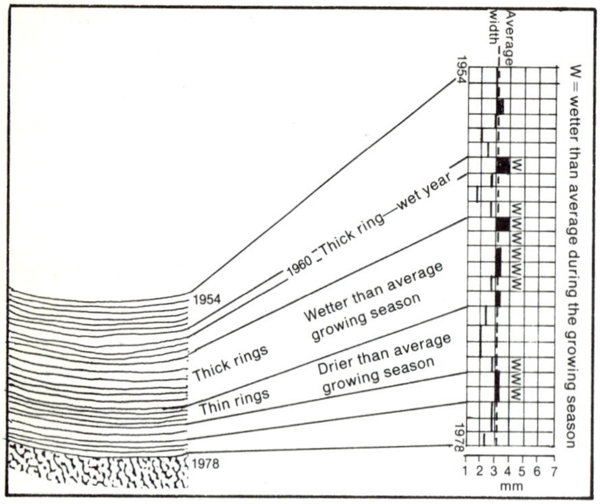

Figure 3.5 Beech tree rings

anomalies in the growth of tree rings, for occasionally a second ring is produced in the same year or a full ring does not develop in all parts of the stem. Standardisation procedures are employed, including calculating **ring-width index values**, this being needed because the width of tree rings decreases with increasing age of the tree. Although such standardisation is useful, at the same time it can blur some of the results, so that slow, long-term climatic changes may be difficult to identify. Other factors than climate are responsible for the nature of the tree rings, so samples are taken, where possible, from specimens from extreme sites and where the influence of man has been least. In the United Kingdom, accurate records of climate have been kept only since 1841, but trees up to 150 years old give us a useful record of past climate.

Carbon dating The third method for providing a chronology is through **carbon dating**. It provides absolute dates of particular events, from which other dates can be determined. Carbon-14 is a radioactive isotope of carbon and is present in fixed quantities in all living organisms.

The importance of carbon-14 in dating is that the isotopes decay at a known rate. By comparing the ratio of carbon-14 to carbon-12 (which does not decay) in a living organism, the variation in this ratio in a dead specimen allows the date of death to be determined with only about a 5 percent error margin. Carbon-14 dating and dendrochronology are often employed

Figure 3.6 The changing position of the continents, 180 million years ago to the present time

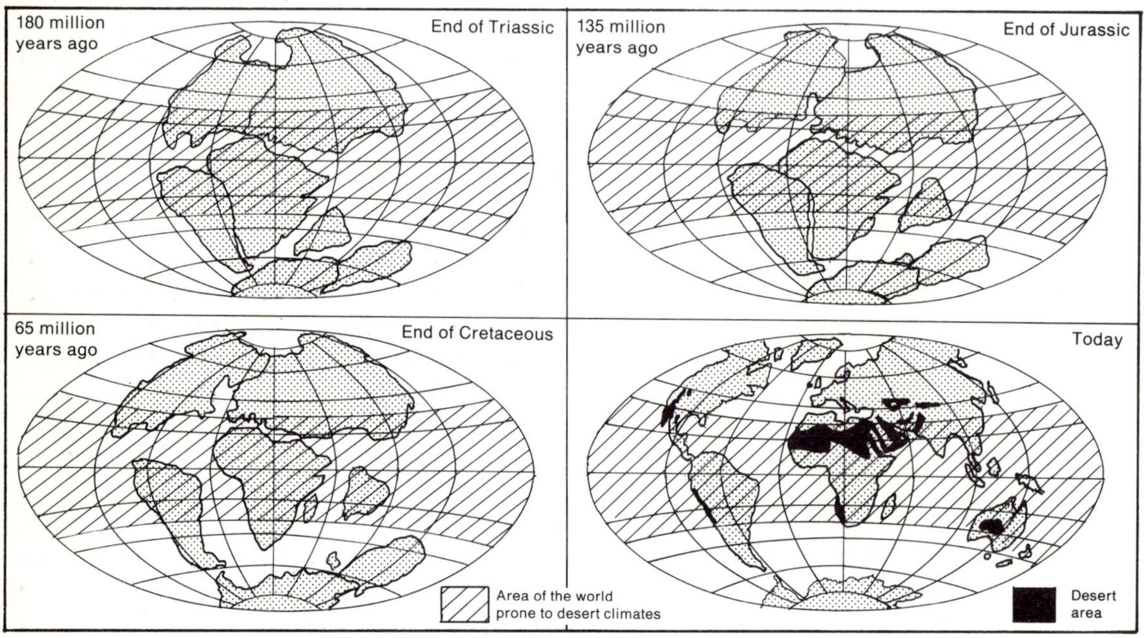

in association to provide checks on the results by each method of analysis.

A further problem in ascertaining geological climatic change is that we now know that the continents have slowly moved their positions and changed their shape and relief. In Permian and Triassic times (180 million years ago), when hot desert conditions prevailed in what we now term Britain, the latitude of the UK would have been very different from what it is now. Figure 3.6 shows the position of the continents on three occasions in the geological past.

Study activity

Discuss the relative advantages and disadvantages of the various techniques available to discover the climate of the geological past.

Historical time

Long before we had instruments available to record and measure the weather, some people found the day-to-day weather such an important part of their lives that they kept diaries, noting the weather in their localities. Annals, chronicles, private and state records, and maritime logs are other sources of **qualitative weather records**. If they are going to be made use of in our quest to reconstruct the climate of the pre-instrumental past, we have to try to ascertain the accuracy and reliability of the documents. Figure 3.7 illustrates the ways in which

Figure 3.7 *Sources of evidence which show the nature of former climates*

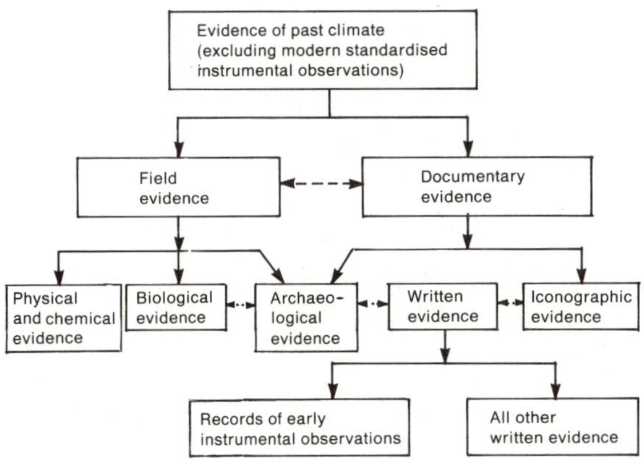

different sources of evidence can be brought together to discover the nature of former climates.

Many of the written sources do not attempt a detailed and accurate account of weather conditions, but include information on weather as part of a more general summary or description. Thus details of harvest yields may include comments on the weather without explaining variations; writers of diaries often described the weather in the context of events they were recording or the clothing they were wearing. The value of the material is also limited by the use of adjectives like 'warm', 'cold', and 'sultry'. The understanding of these terms varies according to individual perception, and can have widely differing interpretations.

Later records include those of meteorological instruments. The barometer and the thermometer were invented in the seventeenth century, and were used at an increasing number of European locations from 1650 onwards. Most of the observers were amateurs, notably doctors and clergy. Later on, official observation records were kept. An early example is the observatory at St Petersburg (now Leningrad) which started to keep records about 1726, printing year-books of weather observations. Gradually an international network developed, leading to the birth of national meteorological societies and resulting in more comprehensive and standardised series of records.

By the eighteenth century efforts were made to provide comparisons of different scales, units and types of thermometers so that some kind of uniformity could be achieved. This was especially true of the location of instruments: the **Glaisher screen** appeared in the 1850s and was replaced in 1880 by the **Stevenson screen** which is still used. The barometer was the first of the instruments to be developed to a satisfactory standard, and so there are good pressure records going back to about 1750. For temperatures, the longest of all series of monthly temperatures is for central England from 1680, while for precipitation, London (Kew) has continuous records dating from 1697 and the Netherlands from 1715.

Study activities

1 Sources of written material are given earlier in this chapter. Describe in more detail the kinds of records that you think would be useful for research into past climates, and compare the advantages and disadvantages of each documentary source.

2 Reconstruction of past climates has led to the theory that there was a little 'Ice Age' in Europe from the latter part of the sixteenth century until the middle of the nineteenth century, when winter temperatures were thought to be much lower than in this century. Some evidence for this theory comes from paintings of the period, especially from the Dutch masters, and from writings, especially those of Charles Dickens. Examine the Christmas cards that are produced showing winter scenes, especially those that are reproductions of early prints or paintings. Analyse the dates of the originals and see if they belong to any particular period. Do you think they accurately reflect the nature of winters in Europe at that time, or are they simply romantic images of a type of Christmas climate that is in fact very rare?

A climatic history of Britain since Roman times

Careful analysis of historical records enables us to reconstruct the climate of past centuries and compare it with the climate today. We shall start this climatic calendar or chronologue at the time when the Romans were in Britain.

1 **The Roman era** It seems Britain was probably chillier and damper than it is today so it is little wonder that Roman soldiers used to a Mediterranean climate complained and found being stationed in Britain a dismal posting.

2 **AD 350–900** Gradually the climate began to improve, with warmer summers and fewer severe winters. Farming began to spread into the uplands.

3 **Early medieval warm epoch, AD 900–1300** This was a period of particularly favourable climate, when conditions were warmer than they had been for several thousand years. Vines were widely established in England, thriving in the warm summers. The Vikings sailed the seas between Britain, Scandinavia, Iceland and Greenland, with little hindrance from storms or from ice in the northern seas.

4 **The Little Ice Age, AD 1400–1850** During the fourteenth century a marked deterioration began which culminated between 1550 and 1800 in a climate with mean annual temperatures at least 1°C below those we experience now. Very severe winters, with snow and frozen rivers, became more frequent while cool, wet summers led to disastrous harvests especially in upland areas, with much hardship for families. There seems to have been a greatly increased variability of climate, with occasional hot, dry summers like that which preceded the Great Fire of London in 1666, interrupting the generally gloomy wet conditions.

5 **The modern period** From about 1850 a steady recovery became evident, culminating between 1896 and 1940 in a period with mild wet winters and warm summers. Since 1940 some cooling has become apparent, with some notably cold winters like those of 1947, 1963 and 1985 but also some very dry, hot summers like those of 1976 and 1983. Although for this period there are ample records, the very multiplicity of data makes the discernment of trends rather difficult.

The major climatic shifts in Britain since 1690 are summarised in Figure 3.8.

Study activity

Table 3.1 lists annual rainfall totals for Edinburgh from 1785 to 1983. From looking at the data series you can see that there is a great year-to-year variability. This exercise attempts to see if there are any trends in the data indicative of climatic change.

Method: Running or moving means are a method of smoothing out the annual variations in order to show the trends of the data more clearly. To calculate a ten-year running mean, find the mean value of rainfall for 1785–94 and plot against 1789, then the mean

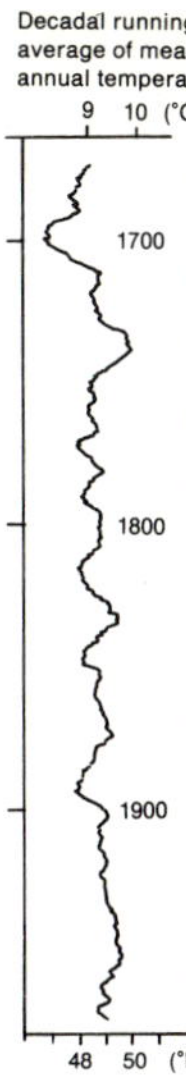

1690s	Extraordinary predominance of cool weather especially from 1692–1702, with late springs, cool summers and cold summers (notably 1695 and 1698). Severe distress in Scotland – 'King William's Dear Years'.
1726–1739	Notably forward springs, warm, dry summers, warm autumns, and a period of good harvests
1740–1748	Generally dry summers, frequent east or north winds. Very severe winter in 1740.
1759–1783	Prevailing warm summers, but frosty winters in 1770s. Wetter 1763–71, with much cold weather in spring.
1799–1820	More east or north winds, tendency for extremes, numerous cool unfavourable seasons. 1799, 1812 and 1816 much distress.
1836–1845	Inclined to cold with wet summers and poor harvests.
1846–1871	Often dry and favourable, especially 1846–47, 1854–59, 1864–71, but 1848 and 1852 very wet and 1860 cold and wet.
1872–1880	Generally rainy, with 1879 extremely bad.
1881–1895	Generally drier, but four severe winters and much coolness in spring.
1896–1939	Period of increasingly westerly weather and of warmer springs. Ben Nevis snowbed disappears in 1933 for the first time since long before 1840.
1940–	Since 1950 a tendency for summers to become cooler, for springs to become later and for rainfall to become more intense in some areas. Increasing flood risk. Mild autumns. Bad winters 1940, 1947 and 1963.

Figure 3.8 *Climatic tendencies in Britain since 1690*

for 1786–95 and plot against 1790, and continue to 1974–83. Try to work in a group and use a calculator or micro-computer to cut down the time taken to do the computation.

Examine the running mean graph. Is there any evidence of change in the mean annual rainfall at Edinburgh? Describe the changes.

What causes climatic change?

While there is still considerable uncertainty as to the degree of climatic change and records are unreliable, it is very difficult to explain the nature and causes of these changes. However a number of theories have been advanced, and in recent years models have been constructed to

Table 3.1 Edinburgh rainfall totals, 1785–1983 (in mm)

1785–90					778.5	594.6	811.5	493.5	741.7	699.0
1791–1800	606.6	939.8	525.8	729.5	929.6	448.3	690.4	573.5	673.3	548.1
1801–10	518.4	541.0	401.6	624.1	411.0	557.8	566.4	738.4	759.5	683.5
1811–20	829.1	688.3	511.8	566.2	554.5	641.1	748.8	548.1	687.3	575.8
1821–30	602.2	663.9	768.4	630.2	561.8	387.9	828.0	641.2	771.2	844.5
1831–40	623.1	590.0	530.3	534.4	640.6	839.0	680.0	788.4	595.6	647.7
1841–50	666.0	428.5	604.5	531.9	676.1	801.1	578.4	777.2	564.1	519.2
1851–60	578.6	800.3	651.0	530.6	516.6	723.4	633.0	618.5	659.6	849.6
1861–70	726.9	861.6	651.0	713.5	600.7	691,6	788.4	725.7	564.6	561.6
1871–80	682.5	989.5	716.0	654.3	618.5	905.5	909.3	632.5	724.4	631.4
1881–90	716.8	767.8	567.2	626.4	446.5	661.7	502.9	631.4	566.7	680.0
1891–1900	615.4	569.2	531.6	717.5	678.7	616.5	578.6	563.9	649.5	795.8
1901–10	576.6	417.6	814.1	610.9	493.5	793.5	766.8	540.5	751.1	699.3
1911–20	537.7	683.5	519.7	578.4	684.8	990.6	536.5	614.2	645.9	698.7
1921–30	628.4	539.6	614.8	726.3	808.4	785.4	874.4	840.0	610.7	804.7
1931–40	809.9	727.1	555.9	687.2	701.7	639.5	736.0	723.4	564.7	706.1
1941–50	637.8	566.3	582.5	748.8	716.4	623.8	658.3	924.3	711.3	733.1
1951–60	786.8	619.1	600.4	892.4	458.1	713.5	650.3	748.0	456.7	617.6
1961–70	642.8	656.5	767.7	503.4	826.5	730.7	636.9	809.1	558.6	599.2
1971–80	565.2	471.6	458.5	528.1	603.4	625.6	782.3	675.6	626.0	672.3
1981–83	610.4	833.8	724.7							

aid the understanding of variations. The causes of climatic change can be divided into two principal categories: external and internal to the climatic system.

External causes

Long-period changes of climate may be caused by astronomical changes. One of the most obvious is the earth's solar heat source. There are two major ways in which variations in the output of solar radiation may occur. The energy output of the sun itself probably varies over time, although there is no observational evidence of any change. Apart from the assumption that there is a fixed solar constant over a time-scale of millions of years, there is also a cycle of sunspot frequency over eleven-year periods. Sunspots are dark areas on the sun implying lower temperatures, and there does seem to be some correlation between sunspot cycles and both temperature and rainfall on some parts of the earth's surface. There are variations in the earth's orbit about the sun which occur in a complex but regular manner. The variations in the shape of the orbit, the tilt of the earth's axis, and the time of year when the earth is closest to the sun may affect the amount of radiation received by the earth.

Internal causes

On the geological time-scale, variations in the surface of the earth that would certainly have resulted in climatic change are the various earth movements. Shifts in the location of continents and oceans, the growth and denudation of mountain ranges would all have had very significant effects. On a much smaller time-scale, the effect of volcanic eruptions which emit large quantities of dust into the atmosphere may have been important, though recent experience from Krakatoa in 1883 and Mt St Helens in 1980 suggests that the effects of a single eruption may last for only a few years. While eruptions can increase the dust content of the atmosphere for a short period, so also can dust from the deserts and from the bare ground around ice sheets. It is possible that dust of this nature results in the cooling of the air because solar radiation is scattered.

In recent times a major source of particles in the atmosphere comes from man's activities. These particles, of urban origin, absorb more radiation and heat the air. More significant than these particles however, is the emission into the atmosphere of carbon dioxide caused through burning fossil fuels. Many climatologists suggest that this may lead to a warming of the earth's climate over the next century.

A third factor in climatic change is variation in the atmospheric circulation. It has been suggested that changes in circulation patterns may be random, but that they do occur is clear. Examination of the climate of the British Isles in the last two hundred years indicates that there are considerable variations in the prevalence of air streams from particular sources. Increases in northerly and easterly winds prolong colder, drier spells of weather, whereas westerly air streams increase the incidence of milder, windier and stormier weather. Figure 3.9 shows the marked differences in the incidence of westerly winds over the last one hundred years.

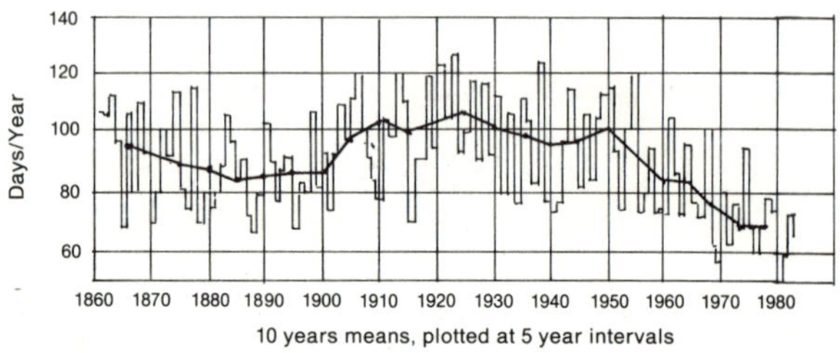

Figure 3.9 Number of days each year with generally westerly winds over the British Isles

Study question

Examine Figure 3.9 showing the change in the frequency of westerly weather types over the British Isles. Is there any relationship between the frequency of this weather type and the annual precipitation at Edinburgh that you investigated in the previous activity?

Short-term climatic changes affecting a few years or decades may be due to feedback effects and complex interactions between the atmosphere, the temperature of the oceans and the amounts of ice in the polar regions. One example is the upwelling of warm waters off the west coast of South America that occasionally displaces the cold Peruvian current which normally flows northwards in this area. These **El Niño** events lead to torrential rainfall in the coastal areas of northern Chile and Peru which normally experience very low annual rainfall totals. The accompanying weakening of the south-easterly trade winds can also lead to periods of unusual or anomalous weather in other parts of the Pacific basin and perhaps in other parts of the globe. During a severe El Niño episode in the early 1980s prolonged droughts affected South Africa and Australia with consequent harvest failures. In Peru itself, the fishing industry—a mainstay of the economy—was severely affected as fish stocks which thrive in the normally cold water conditions of the Peruvian current were severely depleted by the warm water episode. So serious was the climatic disruption during this latest El Niño episode and so great the problems for many countries that requests for help from the developed countries like that in Figure 3.10 appeared in the press.

Figure 3.10 Climatic disruption in the early 1980s resulted in tragedy for millions of people

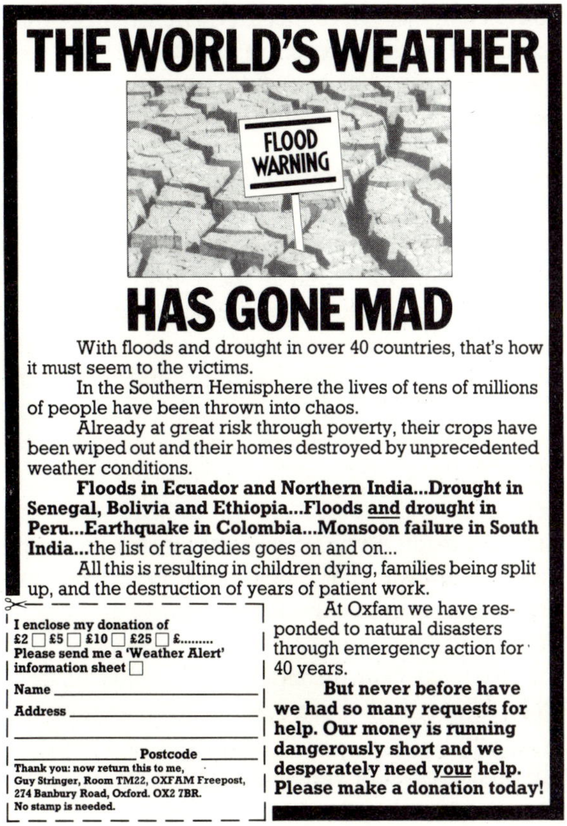

Effects of climatic change

It is difficult to evaluate all the possible effects of climatic change, but two main categories will be examined here: the biological effects and the sociological effects.

Biological effects

A primary influence of climatic change is on plant life. Changes in temperature affect growth and ripening, and the length of the growing season, while changes in precipitation lead to a deficit or surplus of moisture in the soil which can affect plants favourably or adversely. Even small changes in temperature can be significant in terms of the growing season, which is defined as the number of days with temperatures above 6°C. A rise in mean annual temperature over the northern hemisphere from the end of the nineteenth century to about 1940 extended the growing season of plants in England by about two weeks.

Cold-blooded animals are also affected by climatic conditions. Species with stages in their life cycle that include caterpillars and pupae are particularly susceptible to lower temperatures. A decrease in the number of these species has a chain effect working through the ecosystem.

Apart from temperature and precipitation, wind is also an important factor. Changes in wind speed and direction affect the distribution of plants whose seeds are wind-borne, while over the oceans the effects can be seen in the distribution of aquatic life, which has consequences for bird life.

Devastating effects can occur indirectly as a result of climatic change. Conditions can occur that are favourable to the reproduction of pests, viruses and pathogens. Crops can be blighted and animals affected, while people can succumb to diseases and infections. One tragic example of this was the series of warm temperatures and high humidities between 1845 and 1848 in Ireland. Potato blight multiplied and spread, destroying most of the potato crop on which a large proportion of the Irish population was dependent. The result was famine leading to death from starvation. Over one million people died in the great famine.

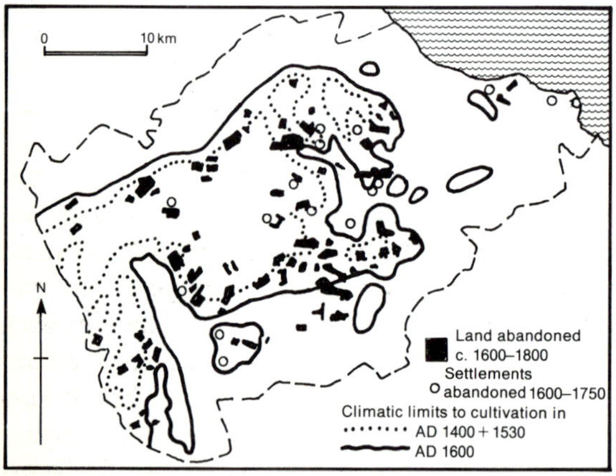

Figure 3.11 Land abandonment and climatic deterioration in part of south-east Scotland, 1600–1800

Study questions

Study Figure 3.11 and answer the following:

1 Why do you think land and settlements were abandoned between 1600 and 1800?

2 Much of the area from which abandonment took place is upland over 300 m. Why were the climatic limits to cultivation different in 1400 than in 1600?

Sociological effects

The degree to which human society is affected by climatic change will depend both on the nature of society and on the type of climatic change. In general, primitive rural societies are most immediately vulnerable, but industrialised societies may have to make longer-term adjustments. While society may be able to adapt to a slow cumulative change, a more sudden change particularly in a climatically marginal environment may prove to be traumatic. For example, if droughts or freezes are either too extreme or too frequent in relation to farm operations then farmers have to make rapid adjustments in land use or even location.

The primary impact of climatic variability on people today is through its influence on food supply. Global food production systems are highly sensitive to even small variations in climate. In the book of Genesis, Joseph warned that seven 'fat' years would be followed by seven 'lean' years and suggested that Pharaoh save up grain in the good years to ensure Egypt's survival in the lean years. Modern society has been slow to learn the lesson, and even today a series of poor crops can easily upset the food–population equation. Climatic changes, whether global or regional in scale, could easily prove to have a destabilising political effect in the future.

Conclusion

The study of climatic change is of value because it sheds light on past climates which have shaped the environment we see about us. Also by examining the climatic behaviour of the past we can learn about possible scenarios for the future. Thus the recent cooling of the climate seems to have been accompanied by more variability from year to year and from season to season, just as seems to have happened in the Little Ice Age. John King, a preacher in Elizabethan times, declared in 1595, 'our years are turned upside down, our summers are no summers'. Very similar sentiments have been expressed in recent years. If the climate in the next hundred years begins to warm up as a result of the rise in carbon-dioxide levels, it may be possible to predict where and to what extent the warming will be most in evidence by considering past epochs of warm climate. A study of the past can be a key to future understanding.

4

Climate, health and well-being

The science of studying the influence of weather and climate on the living organism is term **biometeorology**. In scientific terms it is a new discipline, but from earliest historic times, people have related the influence of weather and climate to a whole range of behaviour patterns and diseases.

In this chapter, some of the results of these studies will be investigated and the relationship between man and his climatic environment will be analysed. The two aspects of the climatic environment which are considered to have the greatest effect on man are temperature and reduced barometric pressure. So what are the problems of human life at high altitude, and in hot and cold climates?

Constraints to human habitation

Altitude

One of the critical effects on people of high altitudes is the amount of oxygen in the atmosphere. Mountaineers have now managed to ascend to the summit of Mt Everest without the use of supplementary oxygen. However, to do that involved a period of adaptation to the reduced oxygen levels of the air at that height. Oxygen is required continuously by the human body, which contains a system to transport oxygen from the atmosphere to the tissues, but has virtually no oxygen reservoirs.

Oxygen deprivation of only a few seconds results in severe effects on the functions of the brain. The body's need for oxygen does not alter with altitude, and severe body stress results from a rapid change of altitude, although a gradual ascent with controlled physical activity does permit a certain amount of adaptation. In recent years, people have learned to cope with reduced pressure at high altitudes by artificial means. Artificial environments are produced in aircraft and spacecraft with pressure-controlled cabins, and mountaineers are able to use oxygen masks and cylinders.

The highest place where people actually live is found in the Andes, up to altitudes of approximately 6000 m. Acclimatisation to these altitudes can take months, even years. Mountain sickness, caused by altitude, includes such symptoms as headache, lassitude, lack of appetite, drowsiness, weakness and breathlessness. These may disappear after a few days if the altitude is not too extreme and the body adjusts to the lower level of oxygen which is the cause of the sickness. When the Olympic Games were held in 1968 in Mexico City, which lies at an altitude of 2305 m, it was feared that the standard of competition would be lower than expected. Most athletes arrived early and allowed plenty of time to adjust to the high-altitude environment. Consequently, the high altitude did not appear to damage the performances of the competitors.

Temperature

Although a much larger number of people live in cold climates than at high altitudes, less is known about the ways in which people adapt to low temperatures. This is partly because people have learned to protect themselves from some of the effects of low temperature. We can increase our insulation by wearing clothes. We can shelter from the environment in dwellings, and we can raise the environmental temperature by the use of fire, or its modern adaptations. Naked people are very vulnerable to low temperatures because like other mammals we have to maintain a constant body temperature of 37°C. If this falls, heat production has to be increased and heat loss reduced in order to maintain that temperature. Shivering is one of the physical responses to increase body temperature. This is inadequate in a cold environment except as a temporary measure, so we are only able to live in this environment because we have produced methods of protection from it and means of altering it to suit us better.

The human race is better adapted for life in a hot climate because our bodies are well supplied with sweat glands, and it is by evaporation of sweat that body temperature is controlled. However, it is necessary to drink plenty of fluid to avoid dehydration in hot climates. People from temperate zones who have never been exposed to tropical conditions can adapt to life in the heat by alterations that take place in their sweating mechanisms. This process occurs over a period of time, and is termed **acclimatisation**.

The effects of radiant energy become more marked with altitude and in addition to the enhanced sunburn risk, skiers can suffer from temporary snow blindness unless they protect their eyes. To avoid sunburn, light-skinned people, particularly those with a fair complexion, should limit their exposure to the sun, gradually increasing the period of exposure once a tan has been achieved. Skin cancer tends to be more prevalent in sunny climates and there is little evidence that acquiring a sun-tan is good for health, despite the money and trouble that many people are prepared to expend to obtain a sun-tanned complexion.

Comfort, discomfort and wind-chill

Whether someone feels comfortable depends not only on temperature but also on factors like humidity and wind speed. Hot, humid conditions with little wind are likely to make a person feel more uncomfortable than the same temperatures experienced in dry conditions. Figure 4.1 shows that while it will feel very cold at temperatures above freezing if the wind is strong, temperatures as low as −10°C can be quite tolerable if the winds are only light.

The cooling power of the wind is known as **wind-chill**, and the concept of combining temperature and wind speed to derive an index was worked out from experiments conducted in Antarctica in the 1940s. The average wind-chill over the globe in January is shown in Figure

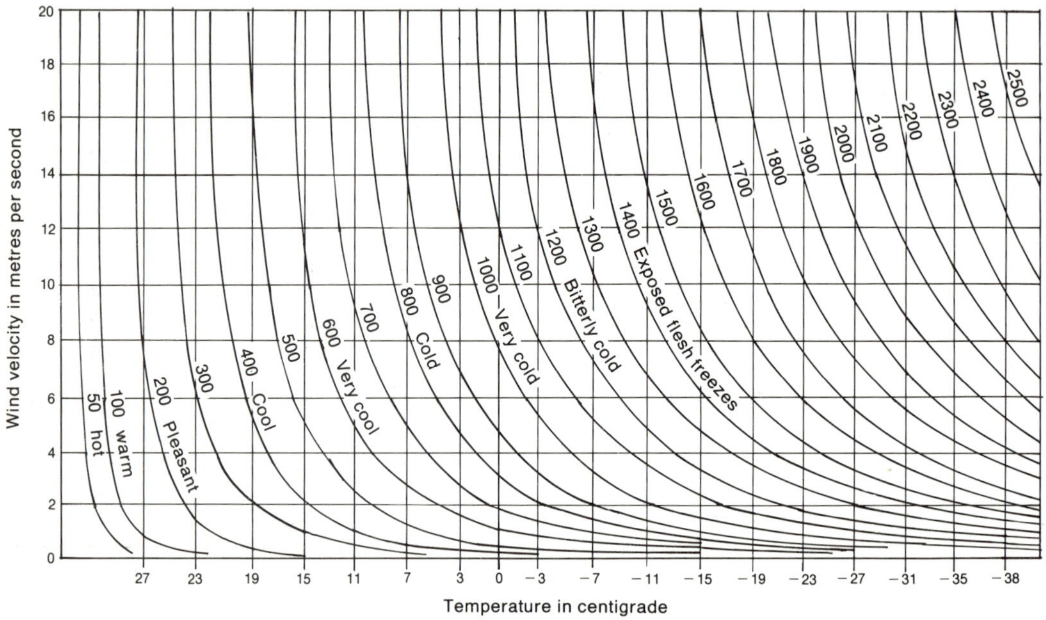

Figure 4.1 Nomogram of dry shade atmospheric cooling

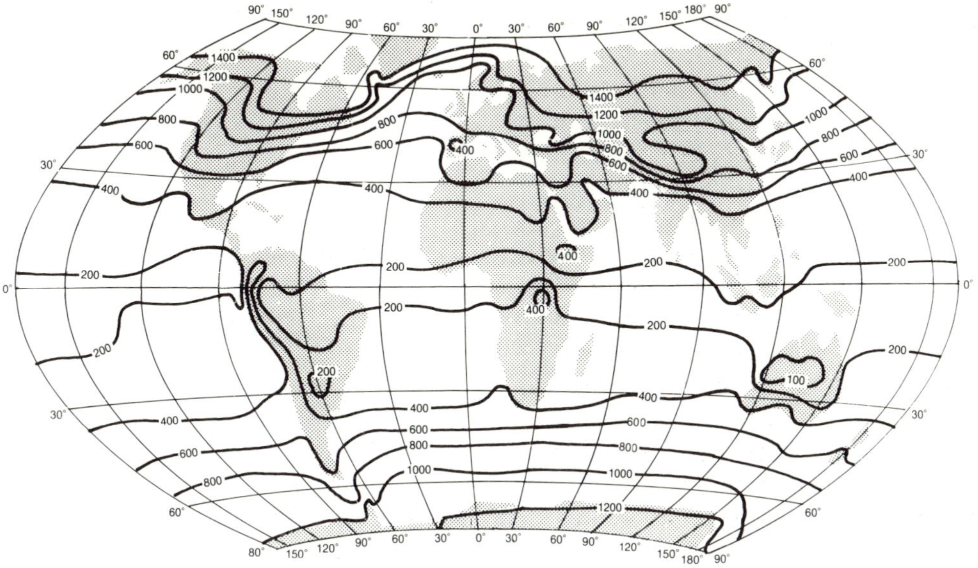

Figure 4.2 *Average wind-chill over the globe in January*

4.2. The lowest values occur over Australia where high temperatures and low wind speeds are found, and the highest values are over northern Canada and Siberia.

The degree of discomfort arising from excessive heat is often measured by an index known as **effective temperature**. This index takes account of the dry-bulb and wet-bulb temperatures, and wind speed. The nomogram in Figure 4.4 enables easy estimation to be

Table 4.1 Meteorological observations in a severe spell of winter weather

Station name	Station number	Dry-bulb temp. (°C)	Wind velocity metres/ sec.	Wind-chill kg. cal/□^{2} /hr
Hurn	862	−1	15.5	1160
Heathrow	772	−1	14.0	1150
Mildenhall	578	−1	5.2	950
Gorleston	497	0	12.4	1090
West Raynham	485	−2	12.9	1160
Aberporth	502	−2	16.5	1210
Ross-on-Wye	627	−3	4.1	950
Filton	628	−2	8.8	1100
Chivenor	707	−1	15.5	1160
Mount Batten	827	1	12.9	1060
Scilly	804	3	14.0	
Culdrose	809	1	15.0	
Guernsey	894	2	15.5	
Elmdon	534	−4	7.2	
Watnall	354	−5	6.2	
Shawbury	414	−6	7.8	
Finningley	360	−4	9.3	
Kilnsea	396	−4	12.9	
Tynemouth	262	−3	9.3	
Valley	302	−2	8.8	
Ringway	334	−5	5.7	
Squires Gate	318	−4	8.3	
Ronaldsway	204	−2	6.7	
Eskdalemuir	162	−6	11.4	
Prestwick	135	−4	7.2	
Leuchars	171	−4	8.3	
Dyce	091	−5	11.9	
Lerwick	005	−4	11.4	
Wick	075	−3	13.4	
Sule Skerry	010	−4	19.1	
Cape Wrath	049	−4	17.6	
Stornoway	022	−4	12.9	
Benbecula	022	−4	9.8	
Tiree	100	−3	13.4	
Aldergrove	917	−4	5.2	
Malin	980	−3	10.3	
Birr	965	−4	.5	
Dublin	969	−3	8.3	
Shannon	962	−3	4.1	
Valentia	953	1	9.3	

Study activity

The data in Table 4.1 was recorded at 18.00 hours one February day, during a strong northerly air stream over the British Isles.

1 With the aid of the wind-chill chart (Figure 4.1), complete the wind-chill column, estimating to the nearest 10 units.

2 Refer to the position of the weather stations, shown in Figure 4.3, and plot the dry-bulb temperatures on a map, drawing isotherms at 2°C intervals.

3 Construct a wind-chill map, drawing the isolines of cooling at 100 kg m²hr intervals.

4 Comment on the differences between the two maps that you have drawn. How useful, do you think, is the wind-chill concept?

Study question

The observations given in Table 4.2 were made at Palmyra Island (5° 52′N, 162° 06′W) during two days in March. With the aid of the nomogram (Figure 4.4), determine the effective temperature for each observation and complete the appropriate column.

At what time of day is the effective temperature highest?

Figure 4.3 Stations published in the Daily Weather Report

made. Effective temperature is used to establish thermal sensation by equating prevailing conditions with a set standard, that being the effect of still, saturated air at a given temperature.

Figure 4.4 Nomogram for calculating effective temperature

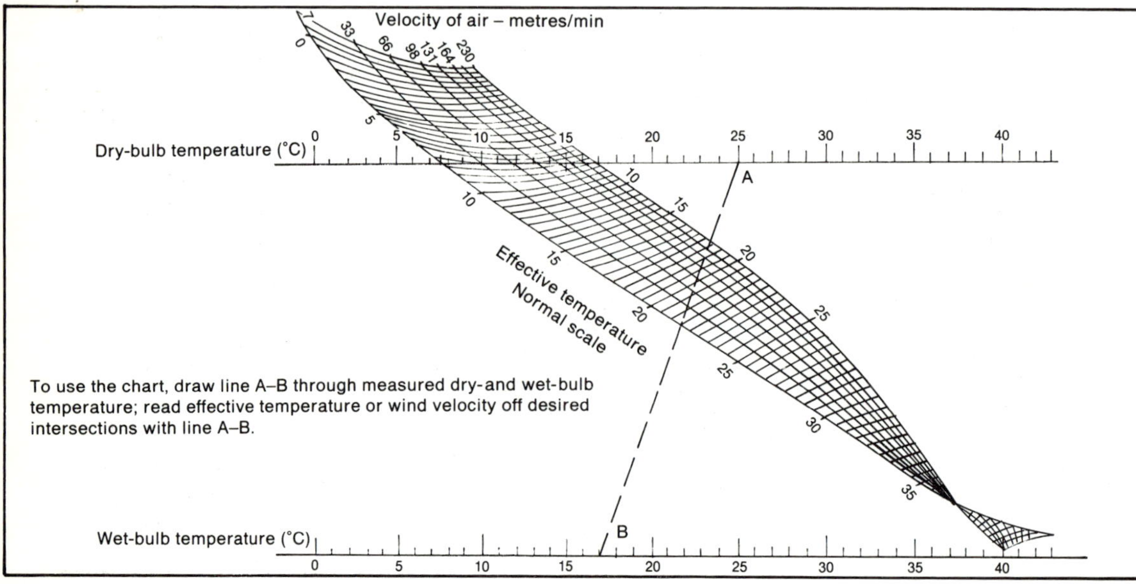

Table 4.2 Meteorological observations at Palmyra Island

		3 March					*18 March*		
Hour (local time)	Dry bulb (°C)	Wet bulb (°C)	Wind speed (m/min)	Effective temp. (°C)	Hour (local time)	Dry bulb (°C)	Wet bulb (°C)	Wind speed (m/min)	Effective temp. (°C)
02	24.2	23.5	53.6		02	27.2	26.5	351.4	
04	24.1	23.4	16		04	27.2	26.0	394.4	
06	24.5	24.2	61.6		06	27.4	26.3	429	
08	25.6	23.8	88.4		08	27.4	25.3	381	
10	27.1	24.6	91		10	28.2	26.0	448	
12	27.8	24.9	107.3		12	28.3	25.7	466.6	
14	24.8	24.1	190.5		14	28.7	25.8	432	
16	27.2	25.2	126.2		16	28.7	26.3	442.6	
18	26.6	24.8	86		18	28.3	25.7	362	
20	24.5	23.9	104.5		20	27.7	26.1	356.6	
22	24.9	23.8	166.4		22	27.3	25.5	335.3	
24	25.6	23.5	279		24	25.4	25.3	485.5	

In a temperate latitude country like Britain there are many more occasions when cold, rather than heat, discomfort is likely to be felt. Consequently the wind-chill index is a more appropriate measure of comfort than the effective temperature index. Average seasonal values have been calculated (see Figure 4.5) to show which parts of the country are generally the most comfortable and those that are most uncomfortable at different times of the year.

Figure 4.5 Seasonal mean wind-chill values for Britain

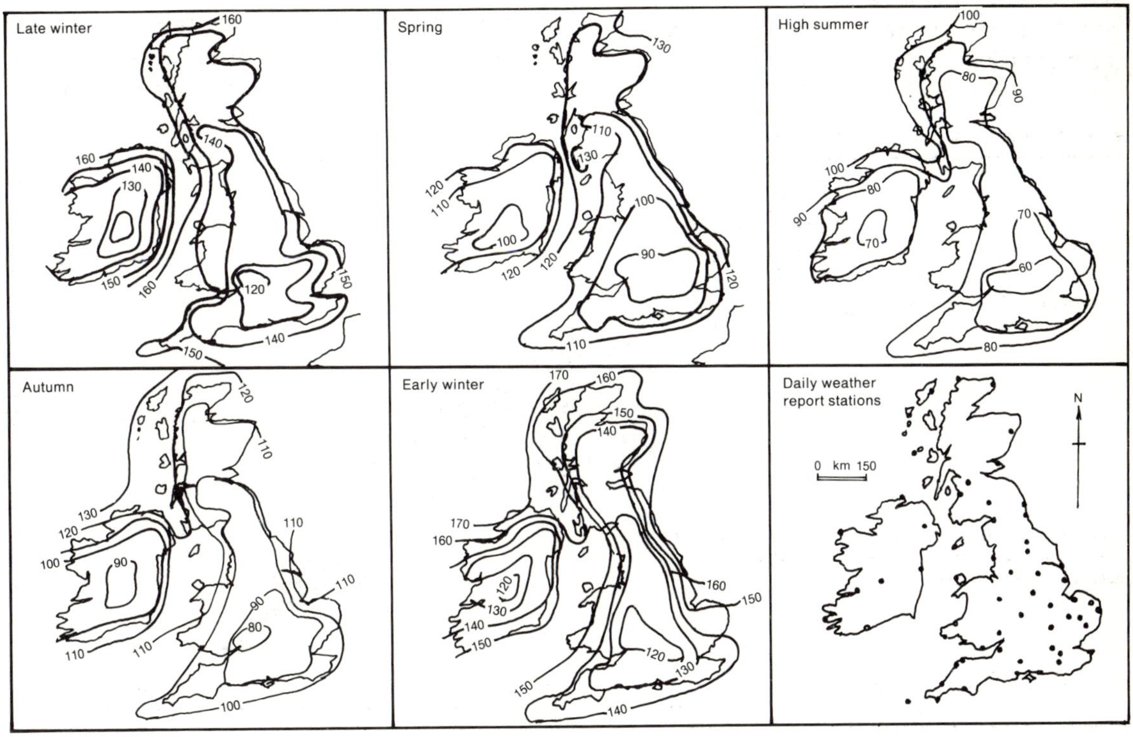

Study questions

Examine the seasonal mean wind-chill maps, (units are calories per square metre per second), based on fifteen years of data, for late winter and high summer (Figure 4.5).

1 In both seasons, lowest values occur over the southern Midlands and central Ireland. Can you explain why?

2 In winter, high values are found in the Hebrides, but winter temperatures are higher there than in many other parts of the British Isles. Why do you think that wind-chill is so high in the Hebrides?

3 Why are wind-chill values higher in both seasons at the coast than inland?

Health and climate

When people are feeling unwell, they often say they are feeling 'under the weather'. Is there any evidence that the weather can affect health?

In answer to that question, it must be said that people of ancient civilisations often considered diseases to be wind-borne. Hippocrates, the father of medicine, wrote a book called *Airs, Waters, Places* in which he blamed the weather for a number of illnesses.

In later times, in many European countries and especially in the USSR and the Alps, climatic health resorts developed often in the mountain areas or beside the sea, on the basis that a stay at such a resort could be very beneficial to health. In the nineteenth century the growth of seaside resorts took place notably because of a belief that the drinking of seawater or bathing in it had curative powers, but also because of the 'salubrious' quality of the sea air. In Britain, people today may recommend 'a breath of sea air' to improve health. Probably the unpolluted air of the sea coasts is a tonic for people more used to breathing the dirtier air of the cities. Around the British coasts, the climate varies from the bracing east-coast weather which makes many people feel active and energetic to the relaxing climate of south Devon and south Cornwall where people may feel quite drowsy.

Until recent times, tuberculosis was a common disease. Diet and cold mountain air were considered to be the treatment for this condition. As a result, many sanatoriums were constructed in Alpine areas like Switzerland, e.g. Leysin.

On the other hand, it can be observed that some weather conditions are hazardous to well-being. Low temperatures in winter lead to a rise in the death-rate among the elderly and infants, especially through hypothermia, which is a condition caused by a severe lowering of body temperature often as the result of living in inadequately heated accommodation. Other causes of death, like bronchitis and pneumonia, are more common when low temperatures reduce the level of the body's resistance to infection. In some places certain winds seem to bring unpleasant medical side-effects. In Alpine countries the very dry, downslope föhn winds are associated with an increase in headaches and migraines. People report that they feel irritable and 'on edge'. Even in Britain, cold and dry northerly or easterly winds produce an increase in the numbers suffering from neuralgia.

At the end of the winter in northern countries like Scandinavia, the population shows signs of tiredness and irritability after the long, dark and cold winters. On the other hand, long periods of hot, sultry weather such as occur in central Europe in summer have been blamed for increases in the number of accidents on the road and at work, probably because such weather tends to reduce concentration. Research suggests that about 30 percent of the population is very sensitive to weather changes.

Climate undoubtedly affects the resistance of the human body to some diseases, and the climate can influence the spread of some diseases, and inhibit the spread of others. Dry, dusty air tends to make the respiratory tract more susceptible to infection. Influenza outbreaks are more common in winter than in summer, partly because people are more susceptible, as noted above. A recent study on the incidence and onset of the common cold in

Derbyshire found that temperature fluctuations were the most significant factor, probably because they make the nasal membranes more susceptible to virus entry. Tropical diseases like malaria and yellow fever are transmitted by a species of mosquito which thrives only in tropical climates.

Weather has to be considered in connection with keeping fit. Jogging and running have recently become very popular. Such exercise can lead to heat exhaustion or collapse much more readily if the temperature is high.

Birth and death-rates and climate

In north-west Europe and in the northern USA, the death-rate is higher in winter than in summer. Respiratory diseases like bronchitis exert considerable stress upon the sick and elderly, as has already been mentioned, while colds and influenza epidemics also take their toll.

However, in Britain there is a saying that 'a green winter fills the churchyard'. This suggests that it is the mild, damp winters that are more damaging to health than the cold, snowy ones. Certainly, in the moister, milder conditions virus infections seem to spread more rapidly and to be more virulent. Conversely, there is a higher mortality rate from accidents and from exposure during the snowy weather. A common cause of heart attacks, among males in particular, is the onset of snowfall in Britain. Men accustomed to travelling to work by car or train and sitting in offices all day attempt to clear snow from paths and drives, thereby overloading their unfit bodies, sometimes leading to an unacceptable level of stress on the heart and circulatory system.

Very hot weather can be as dangerous to the chronically sick as very cold weather. Heat waves, such as sometimes affect American cities in summer, lead to a rise in the death-rate. Often it is the elderly who suffer most, particularly if they cannot afford air conditioning in their homes. Even in Britain, the very hot weather of July 1976 that accompanied the drought that year caused a noticeable rise in hospital admission rates.

Study questions

Examine the data given in Table 4.3.

1 What was the percentage increase in deaths from the week ending 25 June to the week ending 2 July?

2 What is the average weekly number of deaths in Greater London for these summer weeks in 1976? What percentage of the average did the death-rate reach in the week ending 2 July?

3 Calculate the correlation coefficient between mean maximum temperature and mortality in Greater London.

Table 4.3 Mortality in Greater London in the summer of 1976

Week ending	Deaths	Mean maximum temperature
4 June	1377	19.8
11 June	1519	22.1
18 June	1425	23.9
25 June	1475	25.4
2 July	1956	31.6
9 July	1702	31.0
16 July	1425	25.1
23 July	1294	23.2

Conclusion

Weather conditions can affect how we feel, our susceptibility to catching a range of infections and, if we are very young or very old, whether we survive at all. Other nations, particularly in Europe, seem to be more aware than Britain of the importance of climate to our welfare, which is perhaps rather surprising considering our obsession with the weather. Thus Germans, for example, would be quite used to the idea of 'biologically unfavourable weather', such as hot, sultry, close conditions in summer when people feel sleepy, lack concentration and are more prone to accidents.

5

Climatic hazards—the cost of the weather

Costs and damage

Bad weather can cost money and extremes of weather like drought, floods or tropical cyclones frequently cause not only expensive damage but also loss of life. On the global scale, the average annual death toll in the 1970s in disasters was 142 000, but it has steadily risen during the 1980s. At present the average annual death toll from severe weather is probably about one quarter of a million people. Of that total, 40 per cent of the deaths result from floods, 20 per cent from tropical storms and a further 15 per cent from droughts. Climatic changes are not making the earth more dangerous. Instead climatic disasters are acting as triggers which magnify political blunders that have led to world-wide poverty, overpopulation and environmental destruction. Poverty forces people to live on dangerous and unhealthy ground; in Rio de Janeiro steep slopes can slip after heavy rain; in Bangladesh millions of poor people live in river delta areas prone to flooding.

A large percentage of the total deaths occur in underdeveloped Third World countries, where often society is unprepared to cope with a sudden emergency. In a developed middle-latitude country like Britain about 30 people on average die per year as a direct result of the weather, although occasionally disasters like the Lynmouth floods of August 1952 can cause the annual death toll to rise considerably.

A **hazard** can be defined as a potential situation which can cause damage to people, property and the environment. Climatic hazards are a function of both the physical event itself and the nature of the society that is affected. For example, a blizzard in Britain may paralyse transport very quickly but an equivalent amount of snow falling on an American city like Buffalo or an Alpine city like Innsbruck may scarcely be a newsworthy event.

No climatic event must be seen as entirely disastrous. While a flood may damage homes and even cause loss of life, it may also deposit fertile silt on the land and perhaps without the heavy rains that produced the flooding the land

Figure 5.1 *(a) Ski-ing in the Cairngorms, Scotland*

(b) Disruption in the town centre, Hereford

would remain dry and unable to support crops. Figure 5.1 shows two pictures of snow in Britain: in (a) the snow is supporting a ski-ing industry, in (b) it is causing inconvenience in an urban area.

Countries like Britain suffer economic consequences from bad weather, and there have been occasions in recent years when a single widespread severe gale has caused over £50m. damage in a single day. We expect and depend upon rapid and efficient transport links. If there is fog or snow, journeys are delayed or abandoned, goods cannot be distributed and there may be an increase in the number of motor accidents.

Everywhere it is the weakest and poorest sections of society that suffer most in a disaster. The aged, the infirm and the very young are likely to be the most vulnerable, while on an international scale it is the weakest and least robust economies that are most at risk. Hazards then are selective in their impact.

If the frequency of extreme weather is increasing it is likely that hazardous climatic episodes may become more frequent. Figure 5.2 suggests that weather disasters may be becoming more frequent, but the figures must be treated with caution. With improvements in communications more information is circulated throughout the world, so people today may hear and read about disasters that in the past would not have become known.

Figure 5.2 *Global disasters 1947–80*

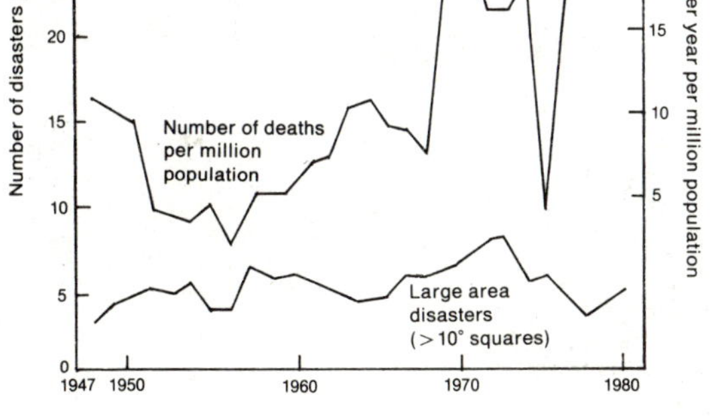

Study activity

Cut out and keep all reports of severe weather from a quality newspaper (e.g. *The Times, Sunday Times, Guardian, Observer, Daily Telegraph*) for one month. At the end of the month, answer the following questions:

1 In which continent have the largest number of severe weather episodes occurred?

2 What type of severe weather has figured most frequently in the reports you have collected?

3 What is the total loss of life, as reported in the paper, from severe weather over the whole earth?

4 Evaluate from actual examples the significance of a hazardous event in the Third World compared with one in northern Europe or North America.

Is it possible to prepare to face the unexpected?

The worst damage and the largest death toll usually occur when an event is unexpected. If it is known that a low-lying area is likely to flood, it is sensible to take precautions—known as making adjustments. The building of houses on that land can be avoided, or if there is a shortage of land, either works of engineering like flood-protection schemes or at the least a system of flood warnings to allow inhabitants time to move valuable possessions upstairs can be undertaken. However, even if a large amount of money is spent, there can be no guarantee of complete protection. It is just too expensive to provide absolute safety. In Britain, most protection schemes are designed to cope with the highest flood that can be expected once in fifty years. If a bigger flood occurs, there will probably be inundation.

Not surprisingly, hazards are best appreciated by those who have experienced them, and it is after such an event that steps will be

Figure 5.3 A dried-up reservoir in Northamptonshire during the 1976 drought

taken to provide protection. Isolated hill farms keep in good stocks of food in the winter because the farmer knows that in a severe winter he may be cut off for weeks at a time. When a very rare event occurs, one that has not been experienced in the lifetime of the inhabitants, there will be a lack of preparedness. This was true of the drought of 1975–6 which occurred in England and Wales but, despite that experience, the drought of 1984 also caused water shortages in Wales. The 1976 drought was the worst since rainfall records began. Many reservoirs dried up and it became necessary in parts of South Wales to ration water by cutting off supplies for up to seventeen hours per day. The design of the water-supply system was not adequate to cope with a deficit of rainfall of such magnitude in a normally wet climate.

A different form of protection is insurance cover. Grain farmers in the prairie provinces of Canada, for instance, fear that their crops may be destroyed by severe hail. A hail insurance scheme gives them peace of mind.

Urbanisation concentrates population and wealth while intensifying land use, and often increases vulnerability. People feel safe from the worst effects of the weather in cities, but this feeling of safety may be quite illusory. When the city of Darwin in Australia was hit by hurricane Tracy on Christmas Day 1974, there was great damage and considerable loss of life. This was partly because houses were not built to withstand storms of such force and, furthermore, the debris flying through the air added greatly to the damage. In London, few people could imagine what a major flood, such as might have occurred before the completion of the Woolwich flood barrier, would be like. Disaster movies may provide a more graphic understanding, but the scenes portrayed are more likely to be considered as fiction than as a possibility in real life. One answer to the problem of increasing people's awareness is to erect posters like the one illustrated in Figure 5.4 showing what could happen if a major tidal surge affected the river Thames.

If the worst happens and there is a disaster, most advanced countries have a well-developed system of aid available. This includes the use of military forces with their specialised equipment (e.g. helicopters) and voluntary organisations. When South Wales was hit by severe blizzards in January 1982, army personnel were eventually sent in to help the local

Figure 5.4 A flood-warning poster issued by the Greater London Council

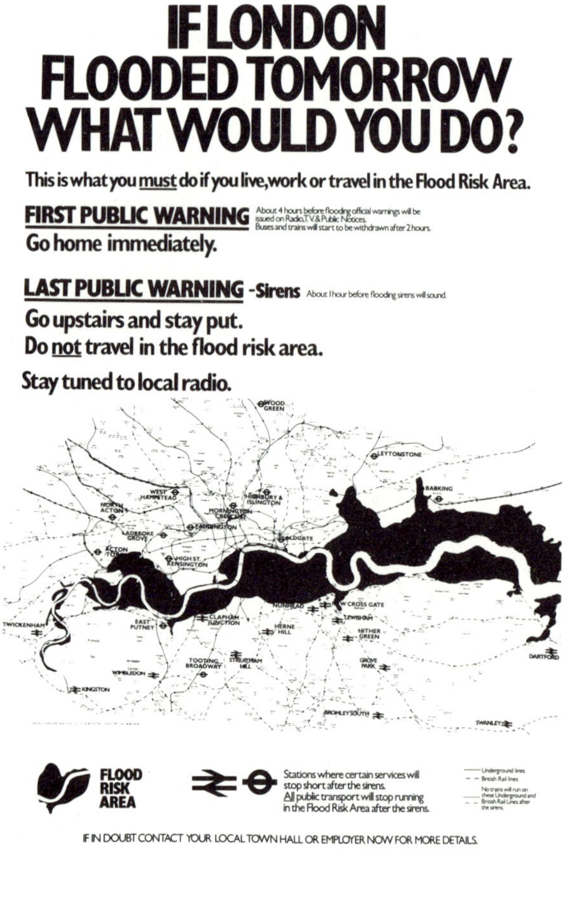

population and supplement local authority services. For some disasters, a voluntary donation scheme is set up to help victims. The European Community has a disaster fund which can give grants to weather-stricken areas to pay for repair work.

Study activity

Make a case for the adequate provision of snow-clearing machinery for a county in southern England. List the main objections that would be likely to be raised in view of the cost of making such provision.

Examples of climatic hazards

Floods

Floods are amongst the most common of all disasters, not only in Britain but also in many other parts of the world. They can occur as the result of a number of events, not only after heavy rain, but during rapid snowmelt, when exceptionally high tides inundate low-lying coastal areas, or when dams or levees collapse. If floods occur as the result of very heavy or persistent rain, their severity will be affected by the character of the river catchment area. If, for example, the catchment is characterised by short, deeply incised valleys, there may be fast-flowing streams and flood peaks will occur very quickly after heavy rain. By contrast, in the flatter country typical of eastern England, rivers flow more sluggishly and so it takes much longer for the flood peak to travel downstream.

Rarely does a year go by without a major flood disaster somewhere in the world, despite the increasing amount of money being spent on flood alleviation works. Since earliest times, people have dwelt on the banks of rivers like the Nile, because the rivers have provided water supplies and easier transport. Historically the worst weather-related catastrophes have been caused by river flooding. Many millions of people were drowned by floods on the Yangtze River in China between 1851 and 1865. Although town builders like the Romans often took care to avoid siting their settlements on the flood plains, the continuing expansion of such towns eventually led to the utilisation of the low-lying areas. Cities like Gloucester, Shrewsbury and Nottingham expanded onto the vulnerable flood plains in the nineteenth and twentieth centuries. Along the East Anglian coast repeated inundations of the North Sea have caused frequent damage to settlements in this region since Saxon times. Northerly winds driving a surge of water into the narrow confines of the North Sea caused havoc in January 1953, the resultant floods claiming hundreds of lives both in England (especially Canvey Island), West Germany and in the Netherlands.

One of the most terrifying flood situations sometimes occurs after an underwater earthquake has generated an enormous wave, or **tsunami**. This can race towards the shore and completely swamp a flat, coastal plain, drowning the inhabitants who have little time to escape inland. In 1976, over 3000 people in the Philippines were killed in this way after an earthquake under the Western Pacific. (See H. Smith's *Oceans and Seas* in this series for further details of this occurrence.)

Study questions

Examine Figure 5.5.

1 Why do you think floods can occur near Hull, Nottingham and London, in areas where the total rainfall amounts are quite low?

2 What kind of flooding would you expect at Fleetwood and Weston Super Mare?

Figure 5.6 shows the front porch of a church in Kings Lynn. Notice the plaques indicating the levels reached by past floods. Many old buildings and bridges have the dates of floods marked on them. See if you can find any in your locality or in an area that you visit. (A visit to your library will help ascertain this.)

3 What date was the highest flood recorded in the photograph?

4 Are there any reports of this flood in old newspapers or local history books?

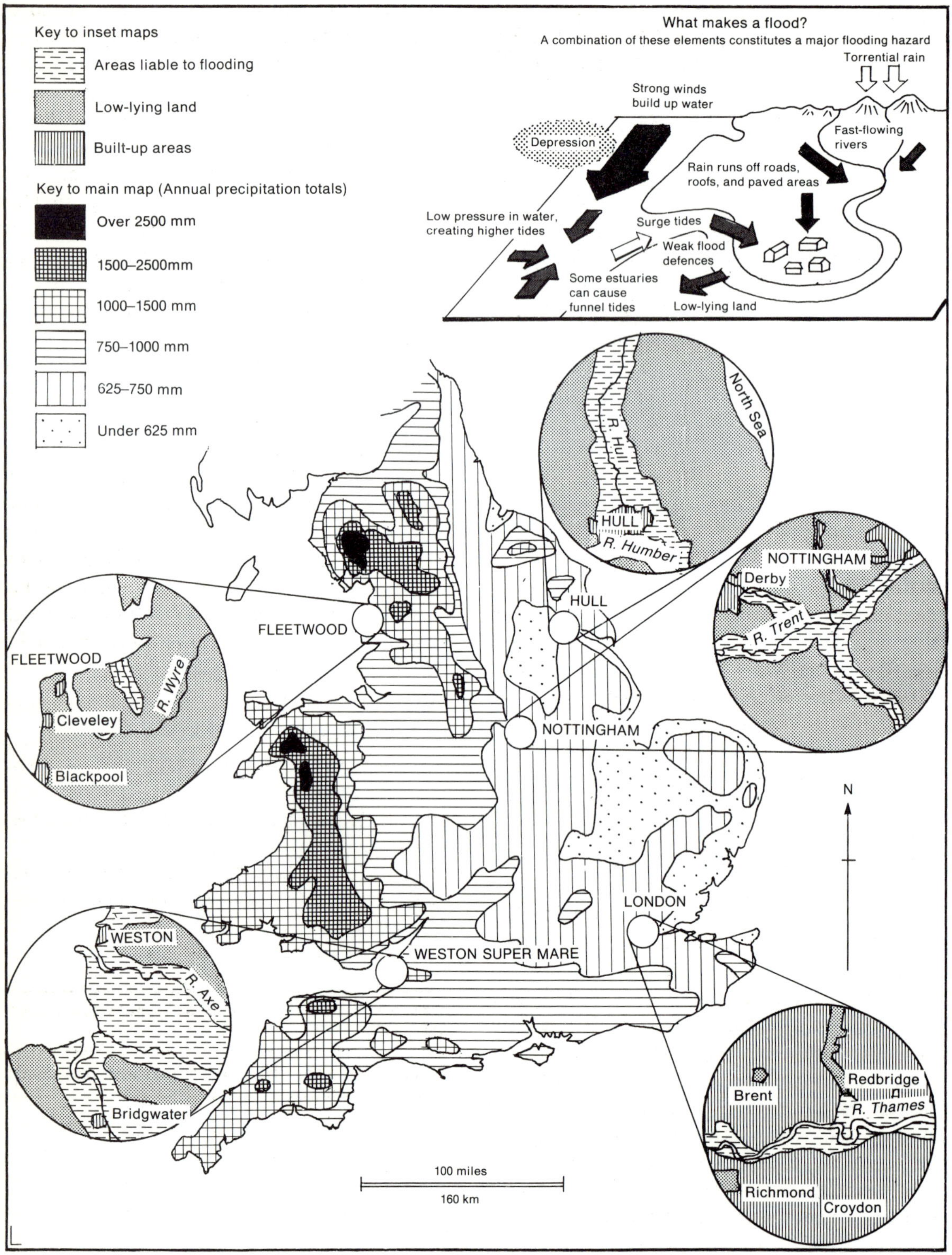

Figure 5.5 *Flooding in England*

Figure 5.6 *The front porch of a church in Kings Lynn*

> 5 Do you think that similar floods could occur in the future?
> 6 Would the local inhabitants have any warning?

Tropical cyclones

A **tropical cyclone** (called a hurricane in certain parts of the world and a typhoon in others) is an intense low-pressure vortex with winds of 33 metres per second—or greater—blowing around it. On average about 80 tropical cyclones occur annually, causing the loss of about 20 000 lives. The most disastrous storms have affected the Bay of Bengal coasts of India and Bangladesh. In one such storm in 1970 over 200 000 people were killed in the Ganges Delta.

Basically, there are two ways in which man can adjust to the threat from these storms. Either he can do nothing and bear the loss, or he can take measures to reduce the potential loss. In less-developed countries there is little alternative to bearing the loss, and hoping that the world community will take pity and help in the aftermath of the disaster.

Storm-warning systems, often based on satellite reconnaisance, have helped to avoid loss of life in regions like the southern USA, where mass evacuation of the cities along the shores of the Gulf of Mexico is sometimes necessary. During the summer of 1985 hurricane Gloria, one of the most severe storms to threaten the United States, slowly approached the Gulf of Mexico coastline. Despite the severity of the storm only two lives were lost since most of the population fled inland to find safety. In Third World countries, adequate warning is often impossible, with many people not having access even to a radio to learn about the advent and progress of a storm. But even with warning, it would be difficult for many people to flee from the oncoming storm on foot or with the aid of animals. In the USA, people can speedily head for safety in their cars, so that damage is likely to be principally to property in the abandoned cities. Even with warning systems, however, there are many areas where little evasive action can be taken. For example, in the West Indies, the populations of the small islands can do little to minimise the effect of an approaching hurricane. Fragile economies continue to be devastated if crops are destroyed, as happened on the island of Dominica in 1979.

> **Study activity**
>
> Table 5.1 gives the annual occurrence of typhoons in Hong Kong in the years from 1884 to 1975. Calculate the mean number of storms per year.
>
> In 1937, 10 000 people died in a typhoon. Do you think such a scale of disaster could happen today?

Table 5.1 Annual occurrence of typhoons in Hong Kong, 1884–1975

Number per year	Number of years with that number
0	31
1	34
2	16
3	7
4	3
5	1
over 5	0

Tornadoes

Tornadoes are violent, whirling vortices of rising air. In many cases the air is so humid that water vapour condenses within the vortex to form a funnel-shaped cloud of water droplets which descends to the earth's surface from the cloud above. Figure 5.7 shows the appearance of a tornado from the ground.

Figure 5.7 A tornado as seen from the ground at Peshawar, Pakistan

The typical tornado damage path is 150 metres wide and up to 10 km long. In the USA, May is the month during which tornadoes are most likely to occur, and it is the central states that are most frequently affected due to the unique combination of weather conditions and topography. In 1974 a single tornado killed 307 people. The National Severe Storm Forecast Centre issues warnings of tornadoes when a direct threat is recognised.

During a tornado the safest place to be is in an underground concrete room or shelter. Most casualties occur as a result of flying debris hitting people caught in the open.

Tornadoes are not unknown in the British Isles, although those experienced in Britain tend to be less intense and less destructive than the fully developed systems that affect North America. The region with the greatest frequency of the more powerful tornadoes in Britain stretches from the North Midlands south-eastwards to Kent. Typically, damage to buildings is very localised, often displaying a linear pattern. Apart from the damage caused by severe horizontal currents, explosive-type damage may occur as well. When a tornado passes close to buildings, the much reduced air pressure can result in windows and even walls bursting outwards.

Study activity

Plot the data given in Table 5.2 in graph form.

1 Suggest reasons as to why the tornado maximum does not occur in the same month as the thunderstorm maximum.

2 When is the tornado season?

Table 5.2 Monthly percentage distribution of tornadoes and thunderstorms in the USA, 1916–64

	J	F	M	A	M	J	J	A	S	O	N	D
Percentage of annual tornadoes	2.1	3.4	9.8	15.7	22.9	18.2	9.8	5.5	4.7	2.5	3.4	2.0
Percentage of annual thunderstorms	1.1	1.6	4.0	7.0	13.3	17.6	21.3	17.8	9.9	3.7	1.6	1.1

Drought

Everyone knows that **drought** is lack of water, but what any particular society defines as a drought varies greatly. For example, in the UK there was a serious drought in the mid-1970s when for eighteen months the rainfall was only about half what is normal. However, the amount of rain that fell during that period was actually greater than that which would be expected in a normal year in a drought-stricken area of the world: the Sahel. The term 'drought' therefore is related to rainfall expectation.

Unlike the sudden impact of a tropical cyclone, the effects of drought grow slowly in an area, and then may persist for months or even years, with the problems of drought gradually increasing their grip on the population. There are several ways of facing up to a drought situation:

1 Do nothing and hope you survive until the drought ends. In primitive societies, the rainmakers will be summoned and various rituals will be performed to induce the rain spirits or rain gods to send the much needed rain.

2 Move away and settle elsewhere. In Third World countries, this often means abandoning the parched countryside for the city, where it is hoped it will be possible to earn a living.

3 Adopt measures to reduce the impact of drought. These might include seeking alternative water supplies. There are well-digging programmes in India, financed by the voluntary agencies; the 'superpowers' may finance prestige engineering works like the construction of major dams (e.g. the Aswan dam); Israel has desalination plants that produce usable water from sea water. Tourism, horticulture and modern society have all raised water demands. Jersey has a desalination plant in use, although neighbouring Guernsey has closed its plant. Different farming styles may be adopted, producing crops that are more drought resistant, reducing the number of animals to conserve the precious vegetation.

Droughts are probably inevitable, but the reactions to them vary. In poorer countries the capital resources are not available to reduce the impact of drought, so it is often necessary for international organisations, like the Red Cross, or the major relief agencies like Save the Children, Oxfam, Christian Aid and Cafod, to bring in food supplies to assist the affected population. In such countries the wealth of a farmer is often measured by the number of livestock owned, and studies in many North African countries have revealed that the land is overstocked in relation to the amount of grazing and fodder available. As the vegetation cover is consumed, there is a spread of desert-like conditions, a process known as **desertification**. The greatest damage is caused by goats. Because they can live in the most marginal environments, they are most common in regions closest to the desert fringe. New tree growth is prevented as goats devour the developing saplings, and many are so sure footed they can climb trees and strip them of their foliage (see Figure 5.8), so destroying the tree.

The creation of the 'Dust Bowl' in the mid-west of the United States in the great drought of the 1930s, so vividly depicted in Steinbeck's novel *The Grapes of Wrath*, indicates that this process can also occur in the developed countries unless careful land-management practices are adopted. Even in Britain, a region at risk is part of East Anglia and the Fens. The wholesale removal of hedges and hedgerows, together with extensive arable farming, exposes the region to the destructive elements of the weather. Dry weather in March, together with

Figure 5.8 The destructive qualities of goats

strong winds, leads to the removal of topsoil exposed by winter ploughing. Already there has been much loss of soil from this region.

A nation's water supply is one of its most important assets but the rainfall which provides it does not arrive in evenly distributed fashion. Nations often respond to the threat of drought by storing grains and other food products. Unfortunately many of the Third World countries are not in a position to do so because they do not produce surpluses, even in good years. Instead they have to turn to the development of products and services that can be exchanged for the precious foreign currency that will enable them to buy food from other countries.

Study activities

1 (a) Compare the ways in which rich and poor countries have been able to combat hurricane damage.
(b) In what ways does hurricane damage cause long-term disruption to the economy? Is your answer the same for rich and poor countries?
2 The drought of the 1980s in the belt of the sub-Saharan countries from Ethiopia in the east to Mali in the west has produced one of the world's worst natural disasters. Amongst the many hundreds of articles written on the subject the following are just two quotations:
(a) 'Instead of taking food to the people it would make more sense to move the people to land where they might be able to grow food and not be perpetually dependent on aid.'
(b) 'The present famine in the Sahelian belt is a manifestation of Nature's way of ridding a superfluous population from areas unable to support it due to the harsh climate. The present rush to save the dying thousands will only act as a temporary palliative. The saved children will grow up, the population will not fall as Nature intends, and the crisis will be renewed inevitably in the near future.'

Do you agree with either of these viewpoints? Give reasons for your reaction to each quotation.

Conclusion

Climatic hazards are potential agents of disaster in terms of what they may do to human settlement or the environment. It is not always easy to identify events as hazards, although certain criteria are usually present. These include loss of life, injury, property damage, economic loss and disruption or even breakdown of social structures and communities. Potentially hazardous atmospheric phenomena occur all over the world, but the impact that they have varies between developing and developed countries, between industrial and agricultural regions, and between densely settled and sparsely settled areas. The variety and relative effectiveness of measures taken to prevent climatic phenomena from causing or resulting in danger is equally various. For the future, warning systems may play a more major role in alerting communities to hazards, and technological advances like improved weather-satellite data and better communications will be of major importance. Modern technology, though, will be to little avail unless the potential victims, who may be illiterate, poor and isolated, can benefit from it. The developed nations enjoy a 'technological cushion' of organised planning and advance-warning systems which alleviate the threat and ease the impact of climatic hazards. In the foreseeable future the trend of an increase in the loss of life from disasters in the developing world seems likely to continue.

6

Econoclimate—the value of the weather

Introduction

Weather and climate can affect both the operation and profitability of an economic venture and the term **econoclimate** is used for studies that examine the interrelationships between weather and economic activities. Many economic, social and political factors influence the production and demand for goods and services, and in many cases climate is just one of the variables that needs to be considered. In the United States, for example, the southern states with their warm sunny weather have provided an attractive location for many new industries, and the area is often referred to as 'the sun belt'. The growth of economic activity in this area is related also to a host of other factors including cost and availability of labour, availability of factory sites, and changes in the distribution of population. Climate is just one term in the equation that has favoured the area.

Some industries, like agriculture, are heavily weather-dependent because the climate limits the range of crops that can be cultivated, and the length of the growing season affects the quality and quantity of crop yields. To be successful, the farmer must take the climatic constraints into consideration, but the decision as to what to produce will also be affected by the demand for particular agricultural products and the price expected to be received for producing them.

Weather influences what people want to buy. The demand for goods like soft drinks or ice-creams and for services like recreational facilities will be heavily influenced by the weather and the entrepreneur needs to be aware of how, and to what extent, changes in the requirements of the customer will be affected by weather changes.

Basic principles

Weather and climate pose a dilemma: we can either accept them, organising our lives and activities around the existing climate, or we can attempt to modify or change the climate. An intermediate response is for us to plan our activities taking into account the weather forecasts, and if necessary to re-schedule our programme if inclement weather seems likely. The farmer contemplating haymaking may well decide to postpone it until a few fine days can be reasonably assured. Sometimes it is even possible to use 'weather protection technology'. In farming, this might mean the adoption of crops that have been developed to withstand drought where there is a strong likelihood of prolonged dry spells in some years.

Climate can be viewed both as a resource and as a restraint. Without an input of solar radiation together with precipitation, crops cannot grow. However, excess rainfall when the crops are at a vulnerable stage of growth may be very damaging.

Regarded as a resource, the atmosphere can be used or misused by man in the same way as other resources. Nitrogen from the atmosphere can be 'fixed' by leguminous plants like peas to improve soil fertility. Conversely, unchecked pollution can render the atmosphere dangerous to life. It has been suggested that one effect of a nuclear war would be to plunge the earth into a long, dark, cold winter as a result of all

Figure 6.1 Traditional method of drying grass in a wet climate

the dust and smoke that would result from atomic explosions, exterminating many plant species, as well as making life almost, if not entirely, intolerable for survivors.

Numerous examples exist to show how modern technology has led to greater rather than to less vulnerability to the weather. For example, a fall in temperature leads to an increased use of electricity and fuels for heating. In the past, an increased demand for gas was quickly met from local sources, but now it requires many hours' advance notice of impending temperature change to increase availability of natural gas from the North Sea. The same vulnerability can be seen in communications: motorways with higher traffic speeds are more dangerous in fog, with the result that climatic surveys of projected routes are often undertaken so that fog-prone areas will be avoided. On an international scale, the business world has become so accustomed to rapid air travel that enormous confusion results when London Heathrow Airport becomes fogbound or, as on one occasion in 1984, New York's Kennedy Airport was affected by a storm that grounded some aircraft and caused others to be diverted, resulting in a chain of delays so that it was four days before Heathrow had caught up on the backlog.

We shall now consider some examples of the 'weather sensitivity' of different sectors of the economy.

The service industries

Tourism and recreation

Weather affects people's choice of a visit to the country or the seaside, the activities engaged in on arrival, and may sometimes even influence safety, particularly in activities such as mountain climbing or sailing. Very often the weather experienced on vacation either makes the holiday enjoyable or spoils the holiday.

Until the nineteenth century, a holiday was a luxury that only the rich could afford, and they often travelled in the belief that 'a change of air' would be beneficial to them. With the advent of the railways, the mass of the population were at least able to leave the cities for day outings. In the years since the early 1950s, the advent of cheap air travel has seen the rise of the package holiday, and by far the most common reason for visiting the Mediterranean beaches is the excellent summer climate experienced there in contrast to the changeable weather of Britain.

A fine summer bank-holiday weekend leads inevitably to traffic jams as city people all make for the same destinations. This 'peaking' of demand leads to problems for those who administer swimming pools, golf courses and the other leisure facilities. Enough staff must be employed to cope with the demand, and plans made and implemented to prevent congestion.

In Britain provision can be made to cope with varieties of weather. The growth of the holiday centre industry (there are five on Hayling Island, Hampshire, alone), with a range of activities for both dry and wet days, represents a good example of 'weather proofing'. Plans for a whole holiday area need to take account of whether there are sufficient wet-weather facilities to entertain visitors, otherwise the area gets a poor reputation and loses trade. Increasingly, the choice of a particular locality for recreation is governed by the perception of the place received from its 'image' on film and television.

Strenuous outdoor sports like sailing, climbing and skiing are particularly weather sensitive. Wise holiday-makers obtain weather forecasts before engaging in these activities. In Scotland, the Cairngorms skiing industry is greatly affected by the amount of snow on the slopes, which is highly variable from year to year, and the force of the winds, with high winds forcing the closure of ski-lifts and tows.

Figure 6.2 Sinah Warren, a holiday centre on Hayling Island, provides a wide range of activities for both dry and wet days all the year round

Key

- A Cabaret/dining room; oak-beamed bar; coffee shop
- B Health and fitness centre; indoor pool; sauna; solarium; gym
- C Disco
- D Games room
- E Sports' field
- F BMX biking
- G Boardsailing; sailing
- H Tennis courts
- I Waterslides

Interest in sports in which the general public can participate has grown rapidly and, despite the climatic problem, skiing in Scotland has grown in popularity so that further expansion seems likely to take place in the future. Centres like Aviemore attempt to compensate for non-skiing days by providing other sports facilities including a heated swimming pool, squash courts and a skating rink.

Figure 6.3 Outdoor recreation activities selected in Canada in relation to daily maximum temperature

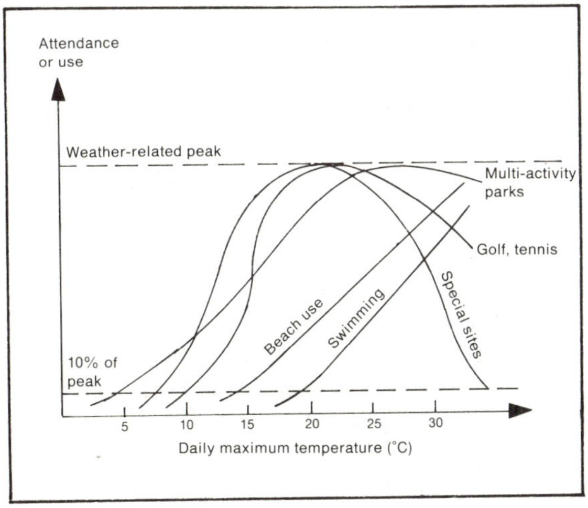

(Other aspects of recreation and tourism are covered by G. J. Ashworth in *Recreation and Tourism* in this series.)

Study activities

1 Choose a sport (e.g. caving, sailing, hand-gliding) and find out what weather conditions are adverse to participation in this sport, and what risks are involved when these conditions occur.

2 Figure 6.3 illustrates the proportion of people selecting different outdoor recreation activities in Canada in relation to daily maximum temperature.

(a) What is the difference in participation in golf and tennis and in swimming?

(b) If the diagram were redrawn with British data, would you expect to see any differences, and if so what would they be and why?

Weather insurance

During a British summer, organisers of outdoor events like county shows and tennis or cricket matches stand to lose a great deal of money if it rains and the events are cancelled or public support is lacking. Weather insurance or

Pluvius insurance (named after the Roman god of rain) can help to limit the losses. Pluvius premiums depend on the date of an event, the location, and the number of hours of cover needed. The insurance company uses the detailed average rainfall records in fixing premiums, which range typically from £12 to £100 (at 1985 rates) for a four-hour fête in the Home Counties of England to about twice that amount for western Scotland. The first drop of rain does not bring the money in, for a sliding scale is applied. Thus if less than one-fiftieth of an inch falls there might be a 20 percent refund. With an abandonment policy, the insurance company will pay up if the event is completely rained off. It is no good glancing out of the window the day before a barbecue and deciding to give the insurance company a ring. Negotiations must be completed at least a fortnight before an event takes place.

Insurance claims for weather damage are highly variable from year to year. In the 1981–2 winter, claims totalled over £5000m., thus making that winter the most costly natural disaster ever to have occurred in Britain.

Retailing

The choice of commodities purchased is influenced by the weather. Clothing is the sector which is most weather sensitive. In a fine, hot summer there is a rush to buy swimwear and light cotton garments, but in a cool, wet summer the shops will be left with large surplus stocks. Apart from the nature of goods purchased, weather also influences overall demand. During cold, wet weather, retail sales are often lower, because people tend to stay indoors, while chain-stores and supermarkets have to try to anticipate the demands of the customer. Date-stamped perishable foods may have to be thrown away if not sold, and while the demand for such foods can soar as the temperature rises, if it is cold people often prefer alternative commodities. Consumption of ice-cream, beer and soft drinks is also weather sensitive. Manufacturers cannot afford to tie up too much capital in the form of reserve stocks, but they need to be prepared for an increased demand in good weather. Not surprisingly, these manufacturers look to the weather forecasters to give them notice of spells of fine weather. Indoor, climate-controlled shopping centres are one solution to the effects of climate on shopper behaviour. Not surprisingly, some of the largest of these centres are found in cities such as Montreal (Place Ville Marie) and Chicago (Schaumberg).

Manufacturing industries

Climate can affect the location of an industry, the demand for its products (as described above) and its ability to transport goods to the market place, as well as such factors as deterioration of stored raw materials and packaging needs of exported goods. Furthermore, the weather can influence the stocks of raw materials, especially in the processing of agricultural products. Until recently, processing could be severely affected by the weather, although artificial modifications are largely available now. Thus the Lancashire cotton industry was aided by the damp climate which reduced the risk of threads snapping during spinning. Today humidifiers ensure the high moisture content of the factories. However, some modifications are very expensive and may attract industry to locations where these are not required, e.g. air-conditioning plants are expensive to build and run, so it is often better to site a factory in a cooler location.

Climate also can act as a deterrent to labour and workers are attracted to the inhospitable parts of Alaska, northern Scandinavia or the Gulf States, for example, only if wages are very much higher than for the same work elsewhere.

Building and construction

There is often a lack of awareness of the huge influence of the climate on the activities of a company, but for industries like building and construction, with much work taking place out-of-doors, the weather cannot be ignored.

In many mid-latitude countries like Britain,

winter puts a brake on much outdoor work in the building industry. Low temperatures inhibit the setting of concrete, and the risk of frost has to be taken into account because of the damage it can cause to construction materials. In 1984–5 the building of the M25 between Wisley and Leatherhead was delayed for three months for this very reason. When there are long, cold, snowy spells, building workers are laid off and have to register as unemployed to obtain state benefits. The annual loss of production exceeds £100m. in Britain.

Rain is probably the most disruptive aspect of climate, affecting operations from initial site surveys, right through the processes to the exterior painting of the finished building. Mention has already been made of low temperatures; high winds are also a problem, impeding work on high buildings and preventing the use of cranes.

Apart from the construction processes, the building industry is very weather sensitive in terms of its product. At the design stage of buildings or engineering works like bridges or motorways, climatic information is required. Constructions have to be able to withstand extremes of climate that may occur in the region in which the construction is taking place. One aspect of research is the use of the wind tunnel to evaluate the ability of the design to withstand high stress through wind force. Structures also have to withstand temperature extremes and the effects of damp through persistent rain.

When a firm tenders for a contract for construction, weather advice is taken. The Meteorological Office in Bracknell, Berkshire, has a service called *CLIMEST* which is intended to give contractors a guide to the amount of adverse weather which may be experienced during the period of a contract. The contractor is then able to use this advice when calculating the length of time the job will take and its cost; and also to plan the best deployment of his labour force, with alternative outdoor and indoor activities depending on the weather. Weather information is available in the afternoons for planning and scheduling the following day's work.

Study activities

1 With reference to your home area, assess the effect of weather on your nearest large shopping centre. Are there any facilities to help combat adverse weather (e.g. covered walkways from car parks to shops, covered shopping precincts, etc.)?
2 Are there any industries in your area that are influenced by the weather? If so, select one and describe the ways in which it can be affected by the weather.
3 Find a building site in your locality, and see if you can arrange an interview with the foreman. Find out what weather conditions cause problems, their effects, and what have been the main delays caused by recent weather in that particular building operation.

Primary industry

When the weather sensitivity of industries throughout the world is ranked (Table 6.1), it can be seen that three primary industries—fishing, agriculture and forestry—all appear in the top five. The most important of the primary economic activities, agriculture, will now be examined.

Table 6.1 Weather influence rank

1	Fishing
2	Agriculture
3	Air transport
4	Forestry
5	Construction

Agriculture

Because world population is doubling about every 36 years, the production of food is vital if the world is not to suffer from an even greater degree of hunger and famine. **Agricultural meteorology** is concerned with the study of weather and climate in relation to agricultural systems and their productivity. The subject en-

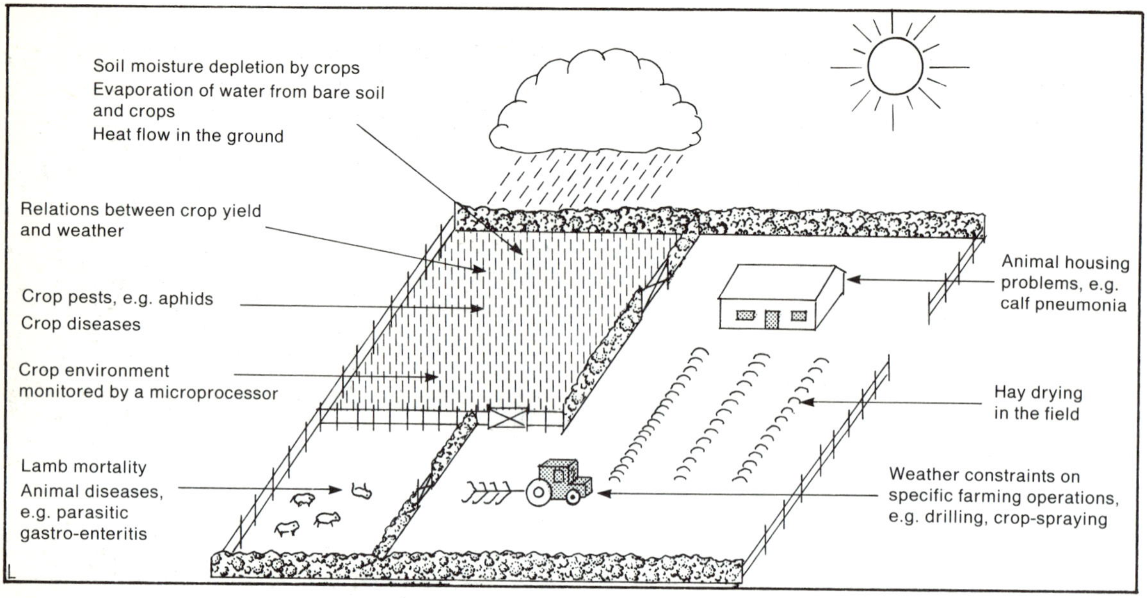

Figure 6.4 *The study of weather and climate in relation to agricultural systems in Britain*

Figure 6.5 *Agricultural regions in Europe*

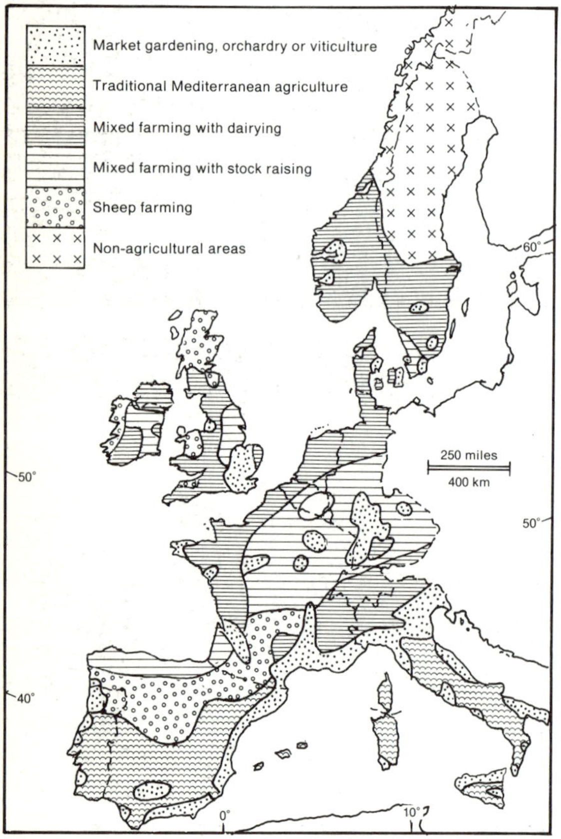

Figure 6.6 *Climatic regions in Europe*

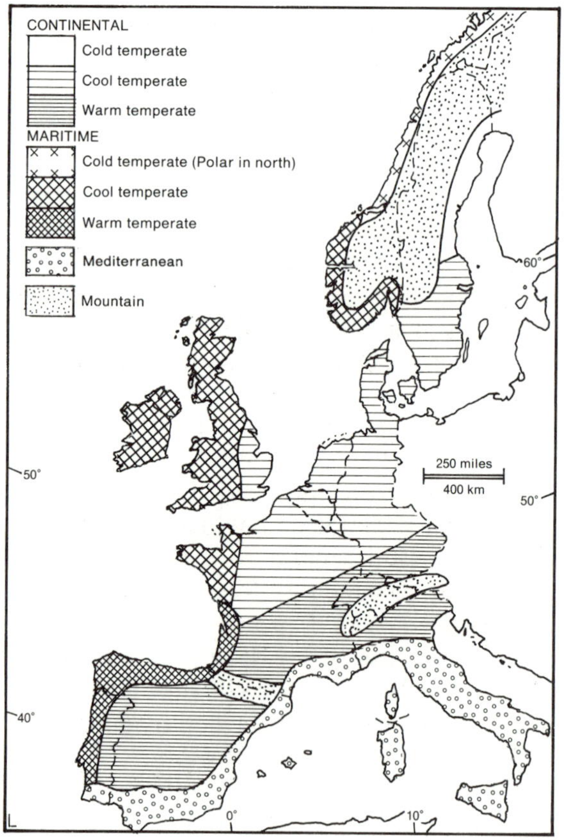

compasses a very wide range both of time and space scales, ranging from the climate of the few millimetres of the atmosphere around a single plant to large-scale assessments of the climatic potential for the agriculture of a region. Figure 6.4 illustrates a few of the studies undertaken to assess how the weather affects agriculture in the United Kingdom.

Weather is one of the major causes of annual variation in agricultural production. Drought during the growing season of the crop is perhaps the most severe problem. Occasionally, as happened in 1972, droughts affect the major cereal-growing regions simultaneously, leading to a shortfall in world production of the major staple crop of wheat. In many regions of the world, irrigation is the main solution to the drought problem, and even in the British Isles, irrigation can greatly increase the quality and quantity of production in most summers, so that farmers find it worth their while to invest in the necessary equipment.

Some agricultural production takes place in artificial environments. Flowers and vegetables are often grown in glasshouses, and some livestock (e.g. chickens) are intensively reared in broiler houses. The horticulturist will require advice on the siting of a glasshouse so that the minimum amount of expensive fuel is used for winter heating. Housed animals must be carefully monitored if disease is not to spread, and a comfortable environment will allow the animal to thrive and to mature more quickly.

Study activities

1 Figure 6.5 is a map of agricultural regions in Europe. Compare the regions with the climatic map in Figure 6.6. How far do you think the climate has influenced the type of farming undertaken?

2 Figure 6.7 is composed of several photographs of the cultivation of weather-sensitive crops. Name the main regions where each crop is cultivated, and assess which climatic features are most important in determining the cultivation of each crop.

Figure 6.7 Cultivation of maize; coffee; grapes; rice

Weather hazards like drought, late frosts and high winds can be particularly damaging: fruit crops like apples are susceptible to frost when the fruit is beginning to form; high winds can flatten a grain crop within a few hours, making it difficult and expensive to harvest. Protection from high winds can be given to crops and animals by erecting shelter belts. These may be lines of trees or artificial structures like walls or netting. You may notice on photographs of hill farms in the uplands of England, Wales and Scotland, that each farm has its belt of trees protecting it from the prevailing winds. In the windy climate of the Isles of Scilly, much of the early flower crop is grown in tiny fields protected from the salt-laden winds by walls. Fruit trees are often planted on south- or west-facing slopes to obtain maximum insolation, while avoiding the valley bottoms where cold air collects and frosts are more likely. In Mediterranean lands and in many of the tropical regions of the Far East, crops are grown on tiny terraces, protecting the plants and the soil from being washed away during heavy rainfall. Hail is a hazard for fruit crops—the Okanagan valley fruit stalls in British Columbia even display specimens of hailstone-impact craters on apples, 1 cm deep and 2 cm in diameter!

Despite recent technological and scientific advances, weather is still the most important variable in agricultural production. Climatic factors have an influence at all stages of the agricultural production chain, from land preparation to harvesting, storing, transporting and processing the crop. With good meteorological advice the farmer can improve his operation, but in many parts of the world the result of lack of advice is very evident. Without diversification, a society practising monoculture will be destitute if the weather does not permit the production of a good crop. Sensible farmers will hedge their losses by having a variety of production so that even if one crop fails, another will survive to ensure some income or food supply.

Apart from direct meteorological information, throughout the world there are agricultural research units investigating the development of strains of crops that are more adaptable and more weather resistant. One national example is the Plant Breeding Station at Aberystwyth, while the Rice Institute in the Philippines gave us 1R8, the green revolution rice. Hardier strains will permit areas that are uncultivated to come into production, will improve the quality of grazing for livestock and resist minor fluctuations in the weather. If world food supplies are to be expanded to keep pace with the increasing population, then it is essential that every effort be made to understand the detailed effects of weather on crops and to put into operation every advance in scientific research that can aid the regular production of food supplies.

Study activities

1 Using suitable examples, discuss the role of climate on agriculture.

2 Identify the regions of the world most susceptible to drought. What kinds of agriculture are practised in such areas? Do you think that the types are in harmony with the climatic environment?

3 If the climate of the British Isles became: (a) drier and warmer; or (b) colder and wetter, how do you think farming types and methods would be affected?

4 If you live in or near the countryside, try to find out from a farmer how weather and climate affect operations on his farm.

7

City climates

Why do towns and cities have their own climates?

Towns and cities have their own climates, distinct and different from the climates in rural areas, and they can be regarded as a modified form of the regional climate. Four out of five British people live in urban areas, while elsewhere cities are growing rapidly, especially in the Third World, thus increasing numbers of the world's population who live in city climates.

What are the characteristics of urban areas that give them a different climate?

Figure 7.1 *Regent's Park and surrounding area in London*

Urban fabric

Urban areas have large surfaces of materials such as concrete, brick and tarmacadam. The thermal properties of these materials are very different from those of grass and crops. In particular they have a **high thermal capacity** which means they absorb and store daytime heat, and release this heat gradually at night. Urban areas are like giant storage heaters and, as we shall see, the result is that cities tend to have higher temperatures, especially at night, than surrounding country areas.

Urban structure

Many cities are composed of tall buildings separated by narrow chasm-like streets. By contrast country areas are much smoother. Look at the photograph (Figure 7.1) and you can see the difference between the park and the surrounding area in London. The rougher surface of the city stirs up air passing across it and a thicker layer of the atmosphere is affected. The whole of the lower atmosphere is affected by the city and the envelope of air that covers the city is called the **urban dome** (see Figure 7.2). You can observe that on a windy day the air will be channelled down some streets, producing uncomfortably gusty conditions, while other areas are sheltered and have almost calm

Figure 7.2 *The 'urban dome'*

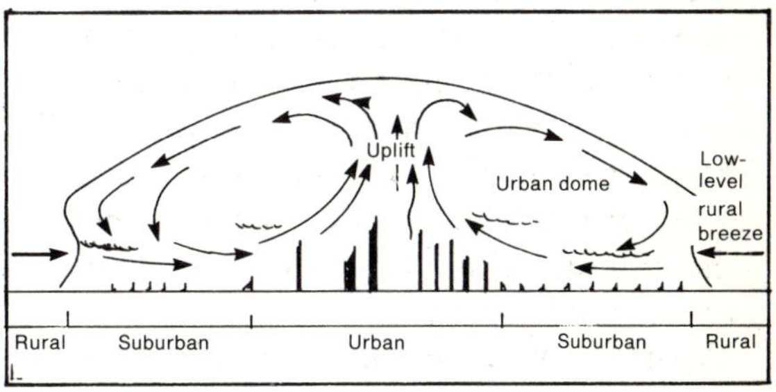

conditions. The building blocks that compose the city can exclude radiation. In a deep city street between tall buildings the ground will be in the shade for long periods of the day.

Waterproofing
Most city surfaces are impermeable, so water that falls runs off quickly into drains and sewers. The result is that the surface dries quickly after rain and the city is like a desert compared with the surrounding countryside where water will remain in puddles for some time. After a snowfall the snow will often be removed from the streets to allow traffic to flow easily, resulting in a situation where the dark surfaces of the city are exposed while snow lies in the country around. At such times the **albedo** or reflective property of the city is very different from the country.

Artificial heat
Motor vehicles and buildings leak heat into the surrounding air and the large numbers of people in the city each produce small amounts of **metabolic heat**. All this extra warmth helps to raise city temperatures. It has been calculated that in New York, on a winter's day, the amount of heat produced artificially by people is $2\frac{1}{2}$ times greater than that received from the sun in the form of solar radiation.

Industrial activities
The industries associated with many towns and cities produce as a by-product a great deal of atmospheric pollution. We shall examine this pollution in more detail in the next chapter, but it is worth noting here that the air in cities tends to be more hazy, and the visibility poorer than is the case in the countryside.

All these changes which accompany urbanization help to produce the distinctive urban climate. Table 7.1 lists the typical changes of climate that the city produces. Even a small cluster of buildings can slightly alter the climate and your own school buildings can be shown to affect the climate to a small extent.

Table 7.1 The effect of an urban area on local climates

Element	Comparison with rural environs
RADIATION	
total on horizontal surface	15–20% less
ultraviolet, winter	30% less
ultraviolet, summer	5% less
CLOUDINESS	
clouds	5–10% more
fog, winter	100% more
fog, summer	30% more
PRECIPITATION	
amounts	5–10% more
days with 5 mm	10% more
TEMPERATURE	
annual mean	0.5–0.8°C more
winter minima	1.0–1.5°C more
RELATIVE HUMIDITY	
annual mean	6% less
winter	2% less
summer	8% less
WIND SPEED	
annual mean	20–30% less
extreme gusts	10–20% less
calms	5–20% more

Study activity

Wait for a sunny, still day and then measure the temperature a few centimetres above: (a) a grass surface; (b) a water surface; and (c) a tarmacadam surface. What differences do you notice? Can you explain them?

Measure the temperature close to buildings and between buildings, as well as at a site away from all buildings, such as on a playing field. Try to draw a map of the temperature around the school. Repeat this exercise at different times of the day. Do the patterns vary? Can you suggest an explanation of your observations?

Study question

'Urban areas have more pleasant weather than rural areas.'

Would you accept, modify or reject this statement? You may refer to weather and climatic studies you have undertaken to illustrate your answer.

The characteristics of the urban climate

Airflow

Between the buildings air movement is very complex because every building affects the flow of air. Under strong wind conditions near tall buildings the wind can become very unpleasant as air is deflected down the windward sides of buildings, forming eddies in the lee of obstacles, or 'jets' through narrow passageways. Architects and builders have to take care when designing new projects that they don't encourage strong winds to blow between buildings and especially if shopping precincts are planned as part of the development (Figure 7.3). For example St George's Walk, Croydon, shown in Figure 7.4, was a wind tunnel until it was covered.

On average winds in towns are less strong than in the country, but they tend to be more gusty. During occasional gales, these gusts can be dangerous to pedestrians, especially old and

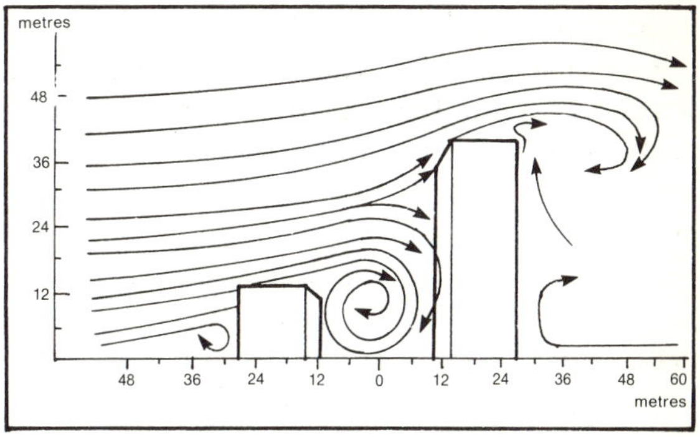

Figure 7.3 Airflow around slab and low building

infirm people. Under calm conditions the city may set up its own wind circulation with air from the country moving into the city, while warm city air rises by the process of **convection**. Such a process is illustrated in Figure 7.2. In the cities of hot climates, such as in the Middle East, it may be possible with careful planning of the streets to encourage cooler air from outside the city to enter and bring a cooling effect to the inhabitants.

Figure 7.4 Due to the tall building immediately behind it, St George's Walk in Croydon (arrowed) was a wind tunnel until it was covered. The lower part of the walk, however, has been left uncovered

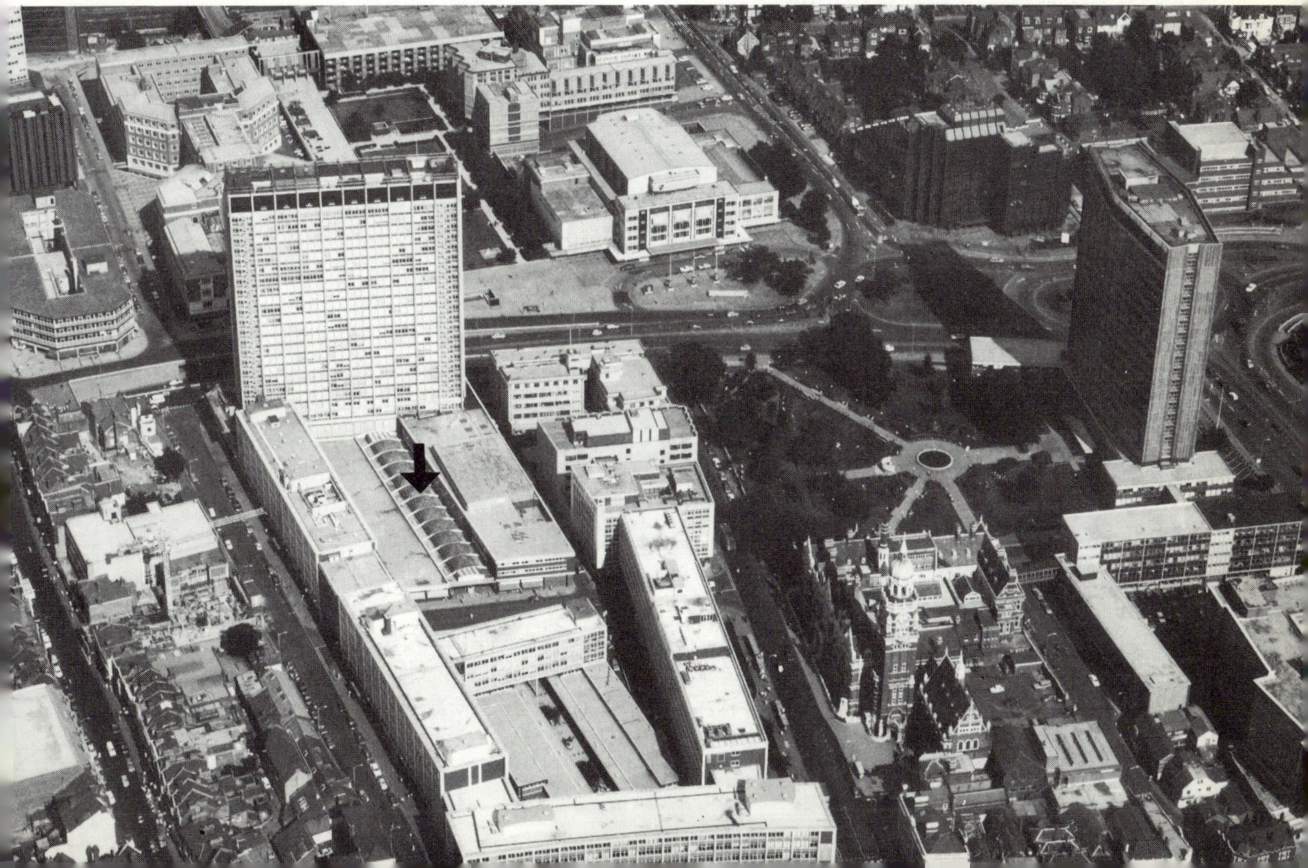

Air temperature

Because there are often insufficient meteorological stations in towns and cities it is only by carrying out special surveys, perhaps using a vehicle with thermometers mounted on it, that we can accurately measure temperature in different parts of the urban area. When such a survey is carried out a map of the air-temperature distribution usually reveals a pattern of isotherm running parallel to the city circumference with increasing warmth towards the city centre. The term **urban heat island** is used to describe such a pattern. The city can be thought of as an island of warmth set in the cooler condition of the surrounding countryside. At the urban–rural boundary temperature rises suddenly upon entering the city, producing a **temperature cliff** (illustrated in Figure 7.5). The strongest or most intense heat islands occur under calm, clear conditions, and in a city like London, the central areas may on occasion be 10–12°C warmer than in the rest of south-east England. In temperate areas this extra city warmth can mean that less artificial heating of buildings is needed, but in tropical climates the city may become stifling and very uncomfortable to live in unless buildings have air conditioning. The temperature distribution in any particular city will be affected by the geography of that city. The presence of water areas and parks, for example, will lead to cooler areas within the city, while the warmest areas are likely to be found where the city land use is high-density housing and industrial or commercial enterprises. In most cities the central areas are likely to be the warmest. You may well wonder if large cities have a more intense heat-island effect than small towns. The evidence from Europe and North America suggests that this is partially true, and this needs to be considered as cities grow in size and merge together, as they have done along the eastern seaboard of the United States.

Figure 7.5 Examples of heat islands

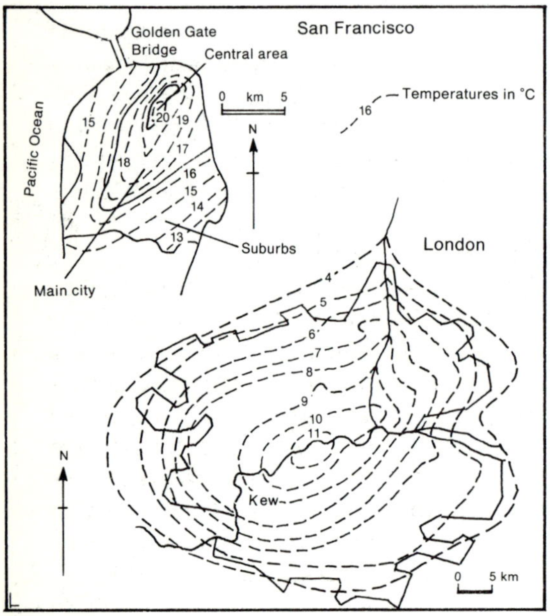

Rainfall

In a Lancashire town in the 1970s it was noticed that more rain fell on weekdays than at weekends. Undoubtedly this was very nice for the local people, but what was the reason? It seems that during the week when all the factories were working, the air was more polluted, and it is known that raindrops need tiny particles to form around. In London over a ten-year period there were about 20 percent more thunderstorms than elsewhere in the south-east. Occasionally a ferocious storm, like that which struck the Hampstead area of London in August 1975, occurs, when in about three hours over 170 mm of rain fell, causing damage estimated at over £1m. On the day there were no other storms over south-east England, and it seems that the warm, built-up area of London helped to trigger the intense storm.

These figures are nowhere near as impressive as those that have been recorded in the little town of La Porte in Indiana, USA, which lies about 50 km downwind of the heavy industrial area around Chicago. At La Porte 31 percent more rainfall and 38 percent more thunderstorms have been recorded than at surrounding stations. Comprehensive work has begun, to check urban precipitation elsewhere in the USA, but the first results of Project Metromex

(Metropolitan Meteorological Experiment) show only a 10 percent precipitation increase in most other cities—perhaps the citizens of La Porte do have something to grumble about.

Sunshine and visibility

The amount of sunshine received in urban areas is less than in the country, especially in the winter months. In London, central parts of the city receive less than half the sunshine hours recorded in the outer suburbs in mid-winter. The reason for this is that urban-smoke pollution acts as a blanket and incoming solar radiation is blocked or scattered by the dustiness of the urban atmosphere. Traditionally, London and many other great cities have had an unenviable reputation for thick fogs: 'Hell is a city much like London, a populous and a smoky city,' wrote Shelley. Most British cities have experienced considerable improvements in visibility and a reduction of pollution levels in recent years and it is good to note that as a result sunshine amounts are increasing.

The most dense fogs are often found in the suburbs rather than in the city centres. This is because the warmer, and often drier, air of the centre helps prevent fog forming, while in the suburbs cooler, damper conditions will favour fog formation on clear, calm nights.

Implications of the urban climate

Very early in human history, we learnt to shelter ourselves against unpleasant weather by constructing buildings. As time progressed and population increased, we became town and city dwellers until today we are so used to living in the controlled climate of the house, the car and the place of work, that we are in danger of forgetting how artificial these environments are. People have accidentally changed atmospheric conditions outside as well as inside buildings, and not all the changes are pleasant. There is a need to understand urban climates so that planning and provision of healthy and pleasant atmospheric environments takes place. By the year 2000 it is estimated that 60 percent of the world's population will live in towns with 5000 or more inhabitants. Cities will devour their surrounding countryside at an ever-increasing pace. One solution might be to encourage 'oases' of quiet and clean air by developing parks and open spaces. Green areas help to prevent flooding by reducing the rate of run-off, and they offer a cool pleasant environment. New cities will need to incorporate urban-climate planning in their designs if they are to provide their inhabitants with the best possible urban climate.

Study activities

1 Test the following hypotheses:
(a) Air temperature increases with proximity to urban and built-up areas.
(b) Relative humidity increases with proximity to urban and built-up areas.

Dry-bulb and wet-bulb temperatures should be measured using a whirling psychrometer or hygrometer, which consists of wet- and dry-bulb thermometers mounted in a wooden frame. This is held manually and whirled in a manner similar to a football rattle for 30 seconds before the readings are taken. Such equipment may be bought relatively cheaply from meteorological instrument manufacturers.

Data collection and fieldwork should be carried out in groups. Try to take measurements at different times of day and under different weather conditions along one or more transect lines across the rural–urban interface. Follow-up work could include the use of Spearman's rank correlation test since the data is discontinuous.

2 Find out if there is a meteorological station in or near your town or city. Ask if you may visit. How representative, in your opinion, is the site of the station of conditions experienced by people living in the city? Find out when observations are taken at the station, and which weather elements are measured. Does your local paper carry any details of weather and climate in your urban area? Do you think the coverage in the paper is adequate?

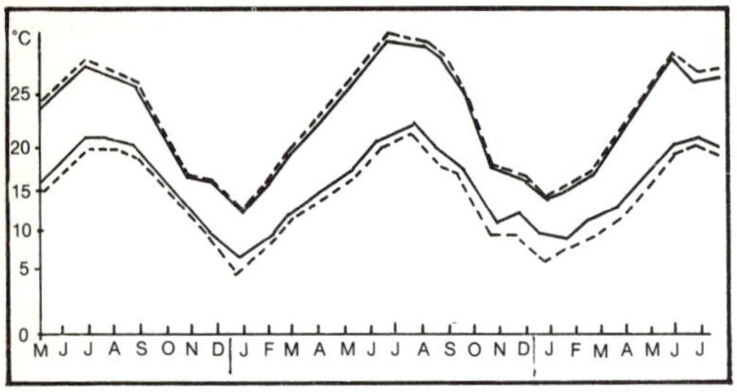

Figure 7.6 *Mean monthly maximum and minimum temperatures recorded at two stations at the same altitude*

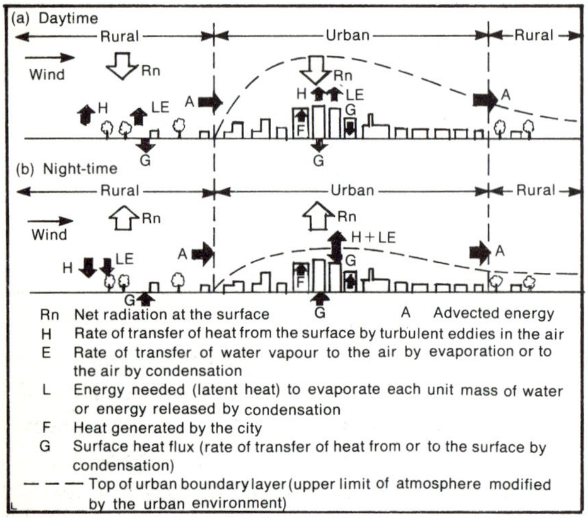

Figure 7.7 *The energy budget of the lower atmosphere and the earth's surface in urban areas and the surrounding rural areas*

Table 7.2 Differences in climate between urban and rural areas

Climate	Differences between urban and rural areas
Air temperature	
(a) Annual mean	0.5 to 1.0°C higher than the surrounding rural areas
(b) Winter minima	1.0 to 2.0°C higher than the surrounding rural areas
Atmospheric humidity	
—relative humidity	2–3% less than surrounding rural areas
Precipitation	
(a) Annual mean	5–10% more than surrounding rural areas
(b) Snowfall	5% less than surrounding rural areas
Cloudiness	
—cloud amount	5–10% more than surrounding rural areas

Study questions

1 Figure 7.6 shows mean monthly maximum and minimum temperatures recorded at two stations at the same altitude. The solid line shows the temperature record at an urban station and the broken line the temperature record at a suburban station. Study the graph and then answer the following questions:

(a) Which station has the greater diurnal range of temperature? Can you suggest reasons to explain why this should be so?

(b) At what times of the year does the suburban maximum temperature exceed the urban maximum temperature?

(c) What evidence of a heat-island effect do the graphs demonstrate?

2 Study Figures 7.7 (a) and (b) which show the energy budget of the lower atmosphere and the earth's surface during the day and the night respectively in urban areas and the rural areas which surround them.

(a) Compare the energy budget of urban areas with that of rural areas during: (i) daytime; (ii) night-time.

(b) Explain why the top of the urban boundary layer is at a lower level at the coldest part of the night than it is during the warmest part of the day.

(c) Using the information given in Figures 7.7 (a) and (b) explain the differences in climate between urban and rural areas illustrated in Table 7.2.

3 (a) Construct a sketch map with isotherms at 0.5°C intervals to show the pattern of temperature distribution you would expect to find over the urban area shown in Figure 7.8 at the coldest part of the night during a winter anti-

Figure 7.8 *Plan of urban area*

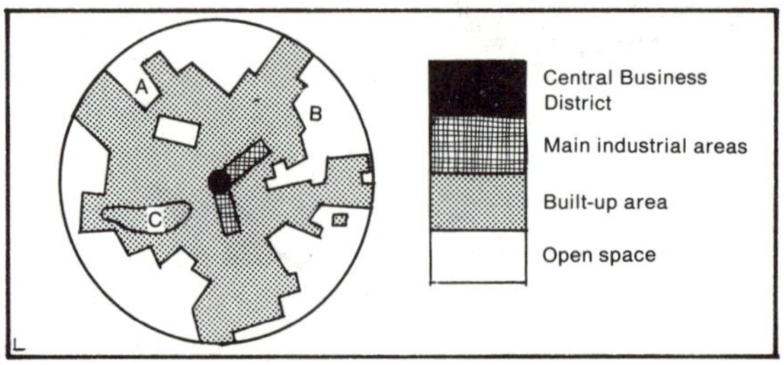

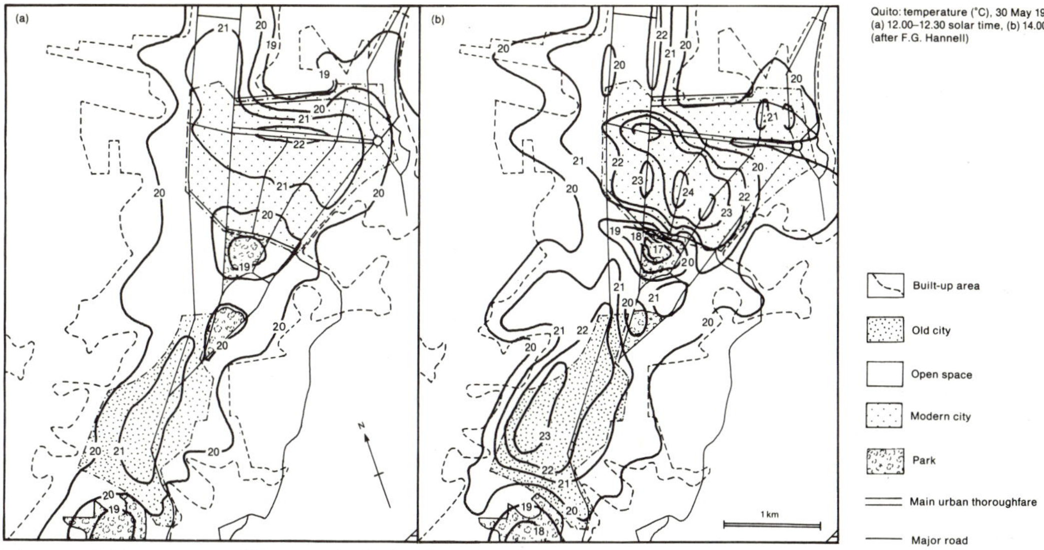

Figure 7.9 Temperature differences in Quito, Ecuador, during one afternoon

cyclone. You should assume that the sky was cloudless and the air calm and that the area in which the town is situated is uniformly flat. The minimum temperature recorded in the area was 2°C and the maximum was 4.5°C.

(b) Attempt to explain the pattern of temperature distribution as shown by the isotherms on your map.

(c) With the aid of a sketch map explain how the pattern of temperature distribution over the area would differ under the same conditions as in question 3(a) if a valley 250 metres deep ran between points A and B marked on Figure 7.8 and the land marked C rose to a height of 250 metres above the general level of the rest of the area.

4 Study the maps in Figures 7.9 (a) and (b) and the following extract, and then answer the questions:

> Quito, the capital of Ecuador, is situated on a plateau at about 3000 metres above sea-level, almost on the Equator. Because its altitude reduces its temperature it has been called the city with the 'eternal spring'. The built-up area [shown on the map] includes some distinctive areas. To the south is the old city consisting of white-painted two to four storey adobe buildings generally interconnected by narrow, unpaved streets, and alleyways less than 3 metres wide. The buildings in the modern city are built of stucco-covered concrete blocks. Many have grassed plots in front or behind, the paved roads are 15 metres wide and away from the main roads, which run more or less in a north–south direction, many houses are detached. The broad east–west avenue is the main axis of the commercial city. Two parks separate the old and modern sections of the city and another park is on a small hill at the southern end of the old city. (Source: *World Weather and Climate* by Riley and Spolton, C.U.P., 1982)

(a) Why, at 12.00–13.30 solar time, is the highest air temperature 1°C higher in the modern city than elsewhere?

(b) What is the probable influence of the small islands of parks on the temperature in the city at various times during the day?

(c) Why would the maximum temperature probably increase by late afternoon?

(d) What is the probable effect of the broad east–west avenue on the local climate?

(e) Briefly describe and explain the main differences in temperature in Quito as a whole, between 12.00 and 15.30 solar time.

8

Atmospheric pollution

What is air pollution?

For a long time industrial cities had a reputation for their smoking chimneys and frequent winter fogs and although there has been a dramatic improvement in Britain in air quality, air pollution has not been totally eliminated. Two factors determine the pollution at a site: the amount of emissions, and the state of the atmosphere. First, then, what are the different types of pollutants?

Fine particles or particulates

These include both solid and liquid particles of dust, grit and visible smoke and are mainly produced by industrial processes such as cement and brick making and by the combustion of fossil fuels like coal and oil both in power stations and in the home. Naturally, the larger particles quickly fall to the ground near to industrial and commercial works, so that it is common to find that near a cement works, for example, the surrounding vegetation has a fine white coating of dust on it. Smaller particles can remain in the atmosphere for many days and contribute to a smoky haze.

Sulphur dioxide

This gas is created when sulphur-bearing fuels such as coal and oil are burnt, and power stations are the major source of sulphur dioxide (SO_2). It is a significant pollutant because it damages vegetation and materials such as building stones by forming a mild acid in combination with rain-water.

Oxides of carbon

The main source of the gas carbon monoxide is vehicle exhausts and in high concentrations the gas can kill. Amounts of carbon monoxide in the air are likely to be highest where traffic is concentrated, as in busy city streets. In Japanese cities it has sometimes been necessary for police on traffic duties to wear breathing apparatus. Carbon dioxide (CO_2) is released wherever fuels are burnt and again concentrations tend to be highest in urban areas.

Hydrocarbons and oxides of nitrogen

Vehicles are also the main source of these pollutants. They are a particular nuisance when they are emitted in large quantities in the presence of strong sunlight because then they are oxidised into a yellow-brown coloured gas, which can irritate the sensitive tissues in noses and throats. This type of pollution is responsible for the smog conditions that can develop in cities like Los Angeles where summers are hot and sunny. Recently high concentrations of this photochemical air pollution have been noted on hot sunny days in summer in southern England and can contribute to hazy conditions.

Of the five major components of pollution in the air, motor vehicles contribute to them all. The Department of the Environment monitors the levels of these substances in the air and makes annual estimates of volumes and the proportions contributed from different sources. According to the most recent statistics (1984 figures), road vehicles contribute about

90 percent of carbon monoxide, about 30 per cent of nitrogen oxides and 14 percent of the hydrocarbons. The major components of air pollution in the UK and their sources are listed in Tables 8.1 and 8.2.

Table 8.1 Major components of air pollution in the UK

Pollutant	Amount in tonnes
Carbon monoxide	8 830 000
Sulphur dioxide	4 040 000
Hydrocarbons	3 290 000
Nitrogen oxide	1 660 000
Smoke	260 000

Table 8.2 Major sources of air pollution in the UK

Source	Quantity (tonnes)	Major pollutant
Petrol engines	8 471 000	Carbon monoxide
Power stations	3 481 000	Sulphur dioxide
Commercial and industrial	2 190 000	Sulphur dioxide
Domestic	990 000	Carbon monoxide
Diesel engines	475 000	Nitrogen oxides
Agricultural	270 000	Carbon monoxide
Railways	67 000	Nitrogen oxides

When a wind is blowing these pollutants are **diffused** (diluted and spread) and transported. On a breezy day try and observe what happens to smoke emerging from a chimney. You will notice how the atmosphere is able to disperse the pollution quite quickly. By contrast on a calm day, it can happen that the temperature in the lower atmosphere increases with height above the ground, instead of decreasing as it usually does. These conditions are described as **inversion conditions** and when they develop the normal turbulent motions in the atmosphere do not occur. The result is that smoke and other pollutants are trapped in the layer of the atmosphere near the ground. Figure 8.1 shows the effect of these very different atmospheric conditions on the smoke. Inversion conditions are most likely to develop in winter, especially at night during calm, clear weather when the air near the ground is chilled by radiative cooling.

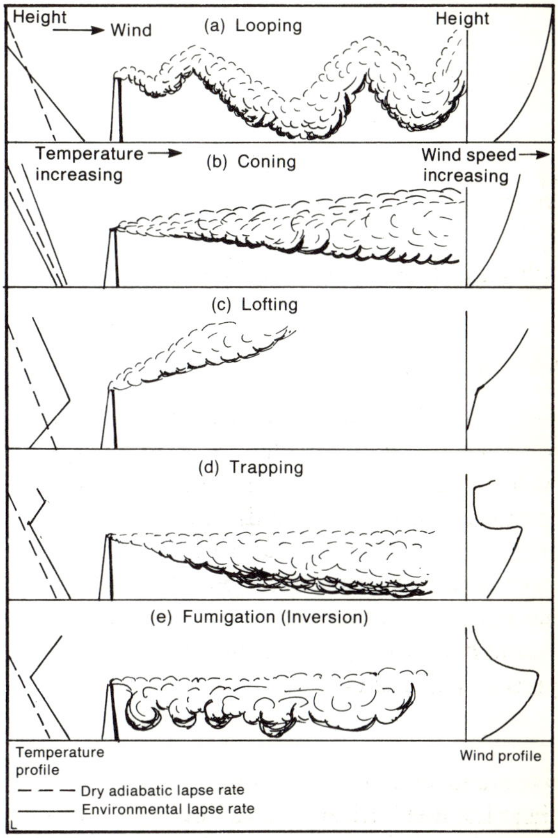

Figure 8.1 Effect of different atmospheric conditions on smoke

Study activity

Keep a record for a week of the behaviour of smoke issuing from a chimney. Try and make your observations (a) early in the morning, for example on the way to school and (b) later in the day, perhaps on the return from school. When making your observations also note the wind speed and direction, cloud amount and temperature. Is there any relationship between smoke behaviour and any of these aspects of weather?

Study questions

1 Study the graph in Figure 8.2 and state the chief source of pollutants shown by the shading in the graph.
2 What differences would you expect to see on a graph of air pollution in the UK?

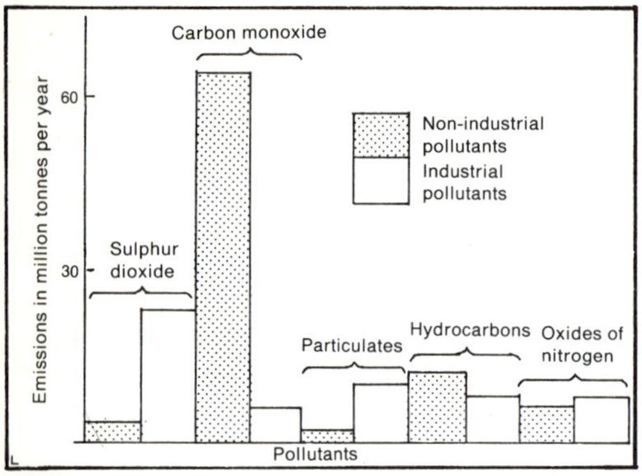

Figure 8.2 *Air pollution in the USA*

Measuring air pollution

At over 1200 sites in the UK air pollution has been regularly measured for periods of up to thirty years. This survey of pollution is coordinated by Warren Spring Laboratory, Stevenage. Local data can sometimes be made available for study, and enquiries (preferably from teachers) should be made to local Environmental Health Officers. For a local area this pollution information can be combined with climatological data from the school or local meteorological station to form the basis of studies in air-pollution climatologies.

Study activities

1 Find out if air pollution is measured in your locality. A leaflet giving details of the National Survey of Air Pollution can be obtained from The Director, Warren Spring Laboratory, PO Box 20, Gunnels Wood Road, Stevenage, Herts SG1 2BX.

2 Draw up a wind-direction pollution rose. These are constructed by averaging pollution concentrations for each wind direction sector for a period of time, e.g. six months. (Further details can be found in D. M. Elsom's text, '1980 Local Air Pollution Studies', *Teaching Geography 6* (Geographical Association), pp. 57–9.)

3 Try to explain why winds from certain directions are more polluted than others.

At official stations pollution is measured using complex equipment, but there is also a simple way to show that air pollution varies spatially. The complex algal plants known as lichens only seem to flourish where the air is clean. They can be found in great profusion on rocks and walls and sometimes festooning tree branches in parts of North Wales and south-west England, but they are much rarer in cities where sulphur dioxide levels are higher.

Study activities

1 Study the map of lichen distribution shown in Figure 8.3. Dark shading illustrates areas of high sulphur dioxide concentrations and no lichens; dots indicate areas of low sulphur dioxide levels and prolific lichen growth.

Write a short account describing the distribution of sulphur dioxide pollution over England and Wales as indicated by lichen abundance. Is there a relationship between the distribution of: (a) industry, and (b) population and sulphur dioxide levels?

2 Try and obtain a book illustrating British lichens (e.g. *Observer's Book of British Lichens*, Frederick Warne). How many species of lichens can you identify in the vicinity of your home? Do you live in an area with high or low sulphur dioxide levels?

Figure 8.3 *Distribution of lichens in Britain*

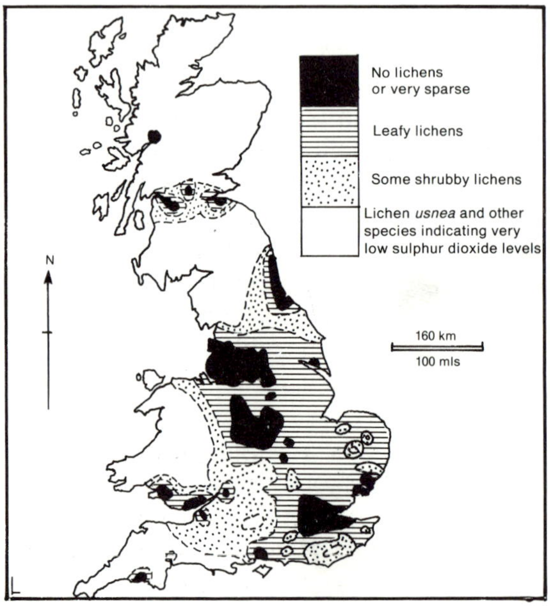

How can we clean up the air?

People have always polluted their environment but with the growth of towns and industry the problem became much more serious. The term 'smog' was used to describe the smoky, 'pea-souper' fogs that sometimes descended on cities and lasted for days. In December 1952, London had four days of smog that caused havoc and led to an increase in the death-rate amongst old people suffering from breathing problems. Something had to be done, and in 1956 the Government passed the Clean Air Act. This banned the emission of dark smoke and allowed local authorities to set up smoke-control districts. The amount of smoke in the atmosphere has declined by about 80 percent since the passing of the Act. Trees and bushes can help to filter the air—and measurements show that a large urban park has cleaner air than the city streets around, but usually controlling air pollution is a costly business, and individual firms have to pay. At Teesside, the company ICI reduced dust output by 98 percent and sulphur dioxide by 97 percent in 1973, costing them £3m. Among the benefits of cleaning up the air are greater varieties of birds and flowers in cities, cleaner buildings and more sunshine.

You will not see much smoke coming from factory chimneys these days, but the Clean Air Act did not legislate against sulphur dioxide pollution, although the levels of this gas near the ground have fallen in recent years. What has happened is that large power stations have been built with tall chimneys. The sulphur dioxide that results from burning coal and oil is released from these chimneys and drifts with the air currents for long distances, dissolving in rain droplets to produce dilute sulphuric acid and this is leading to a worrying international problem.

Sulphuric acid is a waste product of the power stations and one of its effects is to increase the acidity of rain. To cure this problem would involve removing the sulphur from power-station emission which is very expensive and will take many years. Lime is added to rivers and lakes in south-west Scotland in an effort to reduce the acidity. The subject of acid rain will be examined in more detail in the next chapter.

Other atmospheric pollutants are not monitored regularly but are of considerable concern. General emission from motor cars should be reduced in the next few years with the development of 'lean-burn' engines now being introduced. The UK, along with many other European countries, has pledged to phase out the use of lead from petrol in the next decade. Lead has harmful effects on the nervous and brain systems.

Conclusion

We often mistakenly think of pollution as a problem of our own times, forgetting that complaints about the pollution of London's air were common in medieval times. In the Middle Ages, the death penalty could be applied to anyone polluting the capital's air.

People have been using the atmosphere as a rubbish dump for a long time, assuming that it would quickly whisk away all debris dumped into it. We can now appreciate that this policy is misguided and dangerous. Although the quality of the air we breathe has improved, we must not be complacent. While obvious pollution—from smoke, for example—has declined, other problems such as acid rain have arisen. Unfortunately, only a small proportion of British people consider pollution of the environment to be a significant problem—as the more tangible and obnoxious pollution of past centuries has disappeared, we seem to have lost our vigilance and concern.

9

Changing the climate—without really trying

For most of the time that the human race has inhabited the earth the numbers have been small and our technical powers, excluding only the ability to generate fire, have been limited. Thus damage to the environment has been localised. It has only been in the last 200 years that, with increased numbers and technological development, we have been able unintentionally to alter the climate by our activities. The significance of our intervention in the climate system must be viewed against the background of 'natural' variations in climate. Thus if the climate is becoming colder due to 'natural' causes while our activities are having a warming effect, the two trends may cancel one another out.

Even the energy liberated by the explosion of a nuclear weapon is small compared with the energy released by major atmospheric systems like tropical storms. Nevertheless, the climate system is sensitive even to small variations from its original state. The human race is on the verge, if not already capable, of undertaking activities that can disturb the delicate atmospheric energy balance.

Warming us up

World-wide energy use has increased at an average annual rate of about 3.5 percent over the last fifty years. The combined result of population increases and increased per capita energy consumption has been to create enormous urban areas, with extensive industrial sectors where the additional input of energy to the atmosphere is equivalent to the energy from short-wave radiation from the sun. Because the use of energy is disseminated by the countries of the northern hemisphere, the areas where the thermal pollution is likely to be greatest lie in that hemisphere (see Figure 9.1). Regional and global effects of heat emission can be assessed with the help of atmospheric model experiments. They suggest that although the total amount of energy potentially available each year at the earth's surface is more than 20 billion times greater than present energy consumption, waste heat can be locally an important influence on climate.

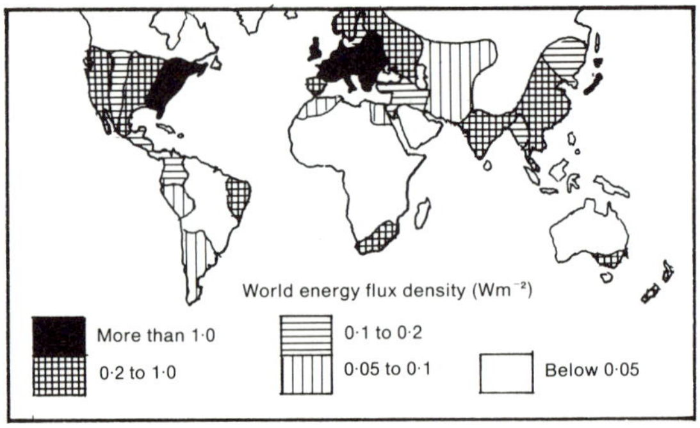

Figure 9.1 Areas where thermal pollution is likely to be greatest

It was the industrial revolution, which began nearly 200 years ago, that required an increase in the use of fossil fuels like coal to power the new machinery. The by-products of combustion include carbon dioxide (CO_2), an invisible gas which is released into the atmosphere. Amounts of carbon dioxide in the atmosphere have been steadily increasing and this situation is giving cause for concern because the whole heat balance of the earth–atmosphere system may be affected. The gas is transparent to solar radiation but absorbs outgoing terrestrial radiation, and thus contributes to the blanket or **greenhouse effect**, warming the lower atmosphere. Between 1860 and 1979 the proportion of carbon dioxide in the atmosphere increased from 290 parts per million to 335 parts per million. Most models predict a doubling of the gas concentration within the next 60–80 years which could lead to an increase in the global mean temperature of 2–3°C. Such an increase could lead to further ice-melt in the polar regions, thereby raising world sea levels. Table 9.1 shows the estimates of change in average surface temperature and precipitation in different latitude belts if atmospheric carbon dioxide increases to 570 parts per million as has been forecast.

Table 9.1 Probable effects of increased carbon dioxide content in the atmosphere

Latitude	Average annual change in surface temperature (°C)	Change in precipitation (percentage)
60°N	+7.5	+18
50°N	+6.0	+ 4
40°N	+6.0	−14
30°N	+4.5	0
20°N	+2.5	+20
10°N	+1.5	+20
Equator	+3.0	0
10°S	+4.0	−20
20°S	+4.5	− 5
30°S	+4.0	+ 5
40°S	+4.0	+12
50°S	+3.0	+12
60°S	+2.5	+12

Agriculture would certainly be affected with the principal grain-growing areas in the northern hemisphere becoming not only warmer but drier. Carbon dioxide is good for crops. It is what they 'breathe in' during photosynthesis and use to build up their tissue. Increased carbon dioxide could, according to some estimates, increase yields of crops like rice and millet by up to 20 percent. Thus there could be some good side-effects in a world with more carbon dioxide.

Predicting future carbon dioxide levels is difficult since we do not know whether the oceans will have the capacity to absorb the extra carbon dioxide that is likely to be placed in the atmosphere. It has been argued that it would be wise to adopt a sane energy policy to protect the threatened climate. Using more of the renewable non-polluting sources of energy and less fossil fuel is a step to be recommended.

Study questions

1 To the average English person, the suggestion that it will get warmer in the future sounds an attractive proposition. Nevertheless, there are serious consequences to human activity if warming were to take place, including the melting of polar ice and the consequent rise in sea level. Suggest some of the adverse consequences that would result from the effects of a warmer atmosphere. (If you are unsure consult D. Ford's *Ice, Land and Man* in this series.)

2 Suggest ways in which the release of carbon dioxide into the atmosphere can be limited and controlled. What are the main difficulties that you think a government would face if it tried to introduce a sensible energy policy?

A dirtier atmosphere

Small particles, known as **aerosols**, are added to the atmosphere by industrial processes, by cars and by many other activities of mankind. There seems to be little doubt that during this century there has been an increase in the rate at which aerosols are produced and the atmosphere has become dirtier. Aerosol particles can scatter and absorb sunlight. The result could be the converse of the effect of carbon dioxide: a reduction in the amount of solar radiation reaching the earth's surface, causing a net cooling of temperature.

An increase of 30 percent per decade in the dustiness of the atmosphere has been recorded at the Maura Loa Observatory in Hawaii, which is far from the sources of pollution. The overall picture of the particle pollution is made more difficult to assess because dustiness in the atmosphere has increased as a result of changes in agricultural practices. The ploughing up of large areas and the removal of tree cover has meant that strong winds are more likely to lift dust into the atmosphere. Evidence of the ability of the atmosphere to transport dust over large distances is sometimes found in Britain. Southerly winds can on occasions transport Saharan dust as far northwards as the British Isles. Motorists complain about the dirtiness of their cars, and when they wash off the dust, a reddish tinge can be seen in the water.

In addition, the dust content of the atmosphere varies according to the number of volcanic eruptions in the recent past. Really dusty eruptions, such as that which occurred in Mexico in 1983, can result in a fine layer of volcanic dust quickly spreading right around the globe under the influence of the upper atmospheric winds. The combined contribution of the 'human' volcano throwing up particles into the atmosphere by human activities and volcanic activity has been to reduce the transparency of the atmosphere during the last one hundred years, as Figure 9.2 shows.

Figure 9.3 Dying trees – the result of air pollution

Rain as acid as vinegar—a major problem of today

Much of the rain that falls on Britain and north-west Europe is acid. This does not mean it will harm you if you are caught in a shower, but it is acid enough to damage plants and fish. **Acid rain** is produced by the burning of fossil fuels. The resulting pollutants are converted in the atmosphere into sulphuric and nitric acids and these harmful substances reach the ground either in dry deposits or mixed with rain (see

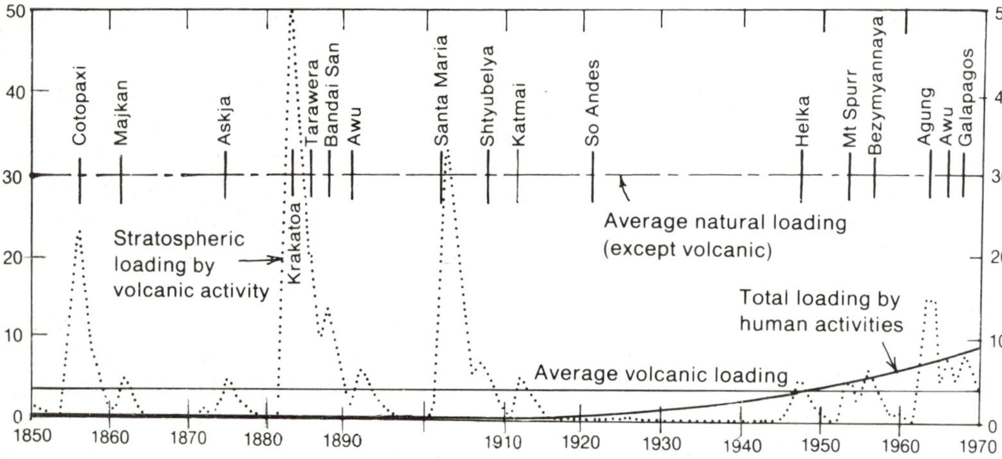

Figure 9.2 The increase in volcanic activity and the amount of aerosols in the atmosphere

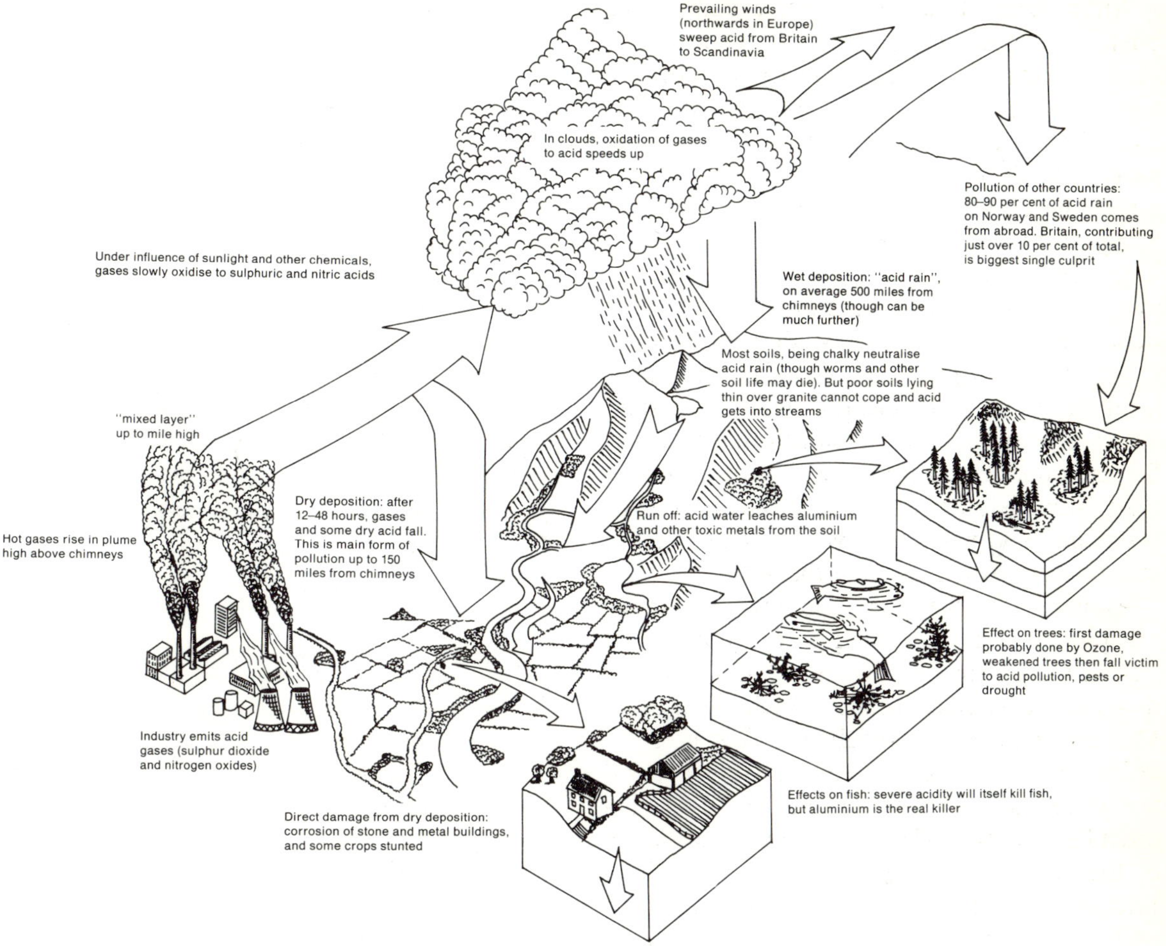

Figure 9.4 *Acid rain: what goes up must come down*

Figure 9.4). The level of acidity is measured on what is called the pH scale. Low pH indicates high acidity, high pH indicates high alkalinity. A pH of 5.6 is the point at which rain is considered acid (see Figure 9.5). At Pitlochry in Scotland rain has fallen with a pH value of 2.4, so the rain was more acid than the vinegar you might put on fish and chips!

Figure 9.5 *The pH scale*

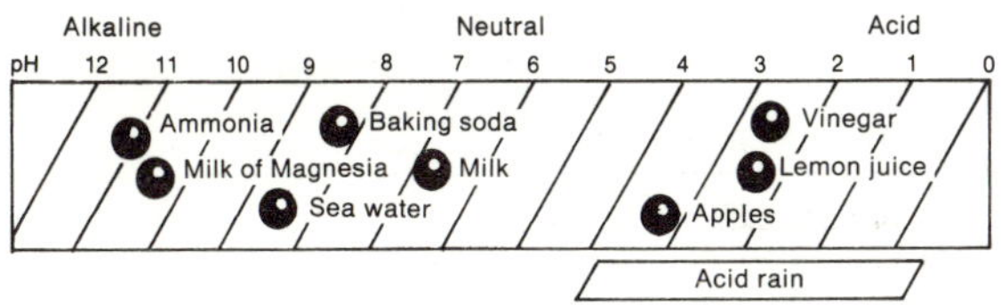

The term **acid rain** was first coined in 1874 but it was not until the 1960s that Norwegian scientists first drew attention to its cumulative environmental effects. In many parts of Germany and Scandinavia the trees are dying and the fish are disappearing from the lakes and rivers. As the rain seeps into the soil it reacts with the acid rocks common in many parts of northern Europe. Aluminium and manganese are released which poison the tree roots. The effects of acid precipitation are governed mainly by the geological conditions: those regions with granite or gneiss bedrock overlain by shallow soils are most susceptible. By the mid-1980s 50 percent of the forest cover of West Germany was affected and in southern

Norway 10 000 sq.km of lakes have lost their fish population. Since the prevailing winds are south-westerly, the emission from power stations in the UK are blamed by our European neighbours. Over 4 million tonnes of sulphur are emitted in Britain and it has been calculated that 60 000 tonnes of that falls on Norway each year.

There is increasing evidence that we in Britain also suffer from the effects of acid rain, particularly in Wales and Scotland. To test this, a major programme called the Acid Drops Project was undertaken in 1985, using 6000 adults and children to measure acidity in the UK. Figure 9.6 shows the levels of acidity measured on two occasions in January 1985. Each map incorporates data from approximately 500 measuring points, with each shaded square having at least one measuring point. On 15 January an easterly airstream off continental Europe brought high acidity levels to many parts of England. Six days later warmer winds from the Atlantic were displacing cold south-easterly winds. The air from the Atlantic was relatively clean, while the south-easterly wind was bringing industrial emissions and 'dirty' air from Europe.

While there is general agreement about what makes rain acid, scientists disagree over the extent of the damage caused by the different chemical components.

Can anything be done about this problem? The Scandinavian governments have called on Britain to reduce drastically its emission of sulphur dioxide into the atmosphere. The technology is available to do this, by installing clearing devices on power station chimneys. However, so far the UK has done very little about it, for two main reasons: (a) the cost involved in purifying emissions would result in even more expensive electricity for the consumer; (b) the Central Electricity Generating Board remains unconvinced that the British power stations are the major culprits.

Figure 9.6 Acidity levels in the UK measured on two occasions in January 1985

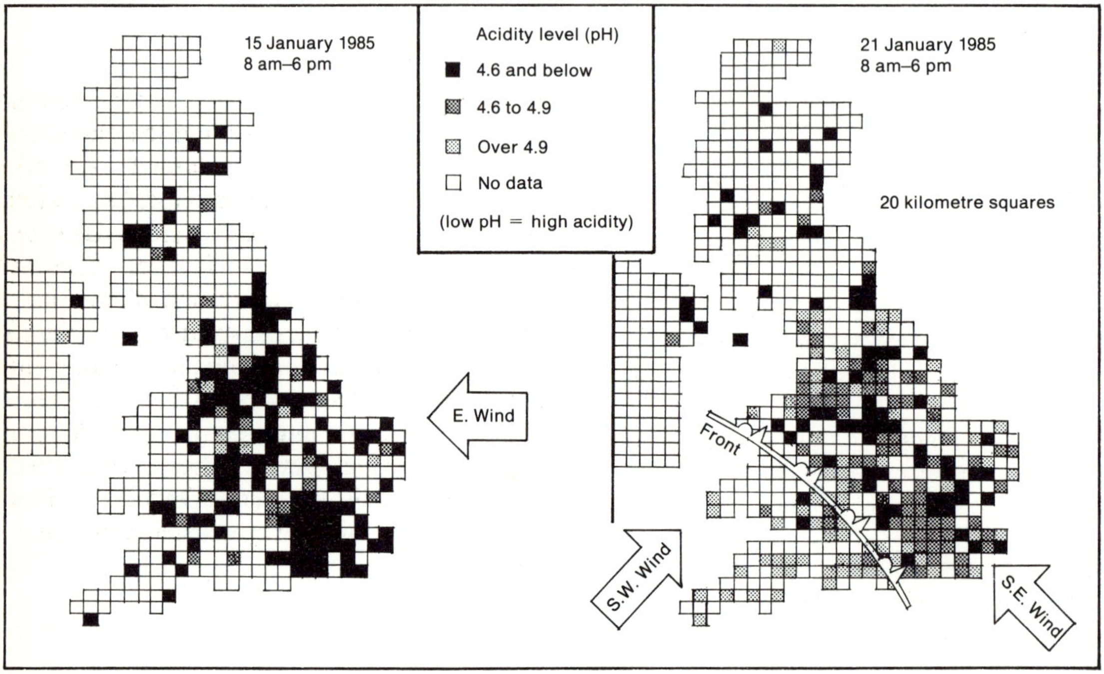

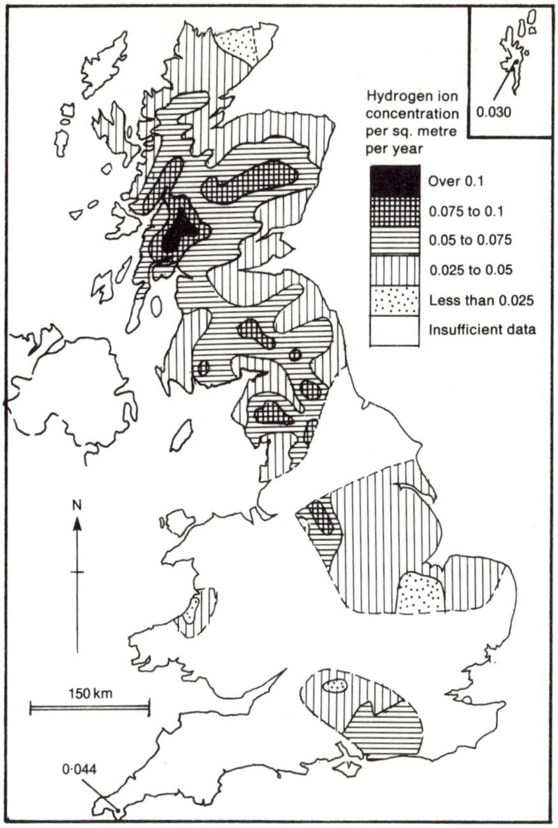

Figure 9.7 *Deposited acidity in Great Britain*

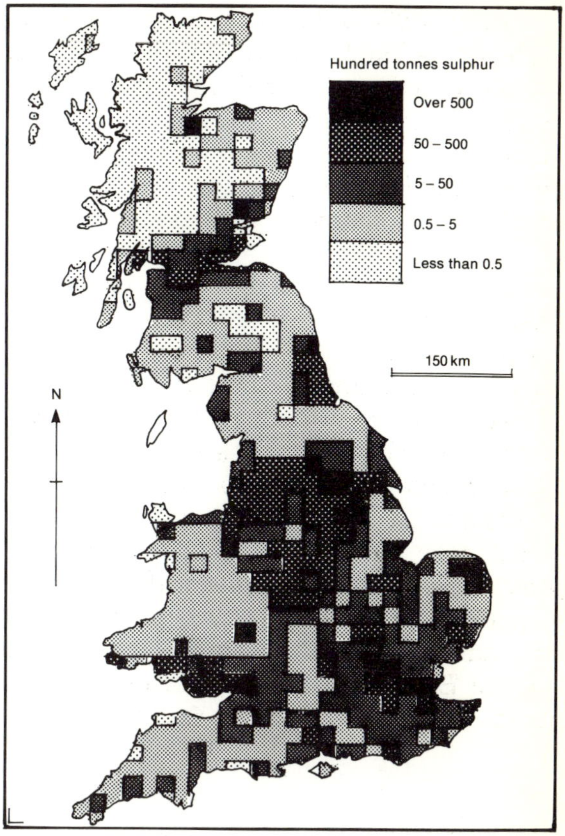

Figure 9.8 *Distribution of sulphur dioxide emissions in Great Britain*

Study questions

1 Figure 9.7 illustrates the deposited acidity in Great Britain.
(a) Describe the distribution shown on the map.
(b) Why do you think values are particularly high in Scotland?
(c) Do you think that electricity consumers in this country would be willing to face higher charges to reduce the acid rain problem in Scandinavia?

2 Figure 9.8 shows the spatial distribution of sulphur dioxide emissions in Britain.
(a) Compare this with Figure 9.7. Why is the distribution different?
(b) Attempt to account for the distribution of: (i) emissions exceeding 5000 tonnes; (ii) emissions exceeding 50 000 tonnes.

Conclusion

Three examples have been given of the way that we are inadvertently changing the climate. In each case the results are already—or soon could be—serious, either on a local or a global scale. Unless action is taken, serious damage to our climatic environment is likely to result, with unpleasant effects for much of mankind. It is one thing to recognise what the problem is, but quite another to change habits and practices so that something is done about it. In each case these are international problems, and it will require joint action by many countries to begin to solve the problems. Only time will tell if the political will can be mustered before it is too late.

10

Weather forecasting and society

In order to exploit the climate we experience, it is necessary for us to have some understanding of the nature of that climate, and in particular to have means of forecasting changes in the weather. Weather forecasting has developed into a largely numerical science, heavily dependent on modern technology, both for taking and recording measurements, as through satellites, and for analysing the information received, which is largely done by computer.

Because of the importance to society of weather forecasting, the processes involved will be examined in some detail.

Weather forecasts

A weather forecast is a statement of the weather expected at a specified time in a stated location. The forecasts for the United Kingdom are prepared by the Meteorological Office in Bracknell, Berkshire. Reports of recent weather, forecasts and warnings of severe conditions in the future are issued from Bracknell through a number of different channels which will be described later.

Figure 10.1 Senior and Assistant Forecasters at work in the Central Forecast Office, Bracknell

Most of the plotting, analysis and forecasting is done by computer supervised by a team of weather forecasters working round the clock. The most significant features of the weather are transcribed into words to be available for general forecasts and for pilots, sailors and farmers. In offices such as the London Weather Centre, other meteorologists work to interpret the general forecast for their own area and for specific needs.

Forecasts are made, using different methods for different periods in the future. The basic principles change only slowly, and much progress was made in the years up to 1970 in the assessment of charts, but since the large-scale use of computers, more detail can be assimilated and prediction can take place further into the future, working on the basic laws of the atmosphere.

Short-period forecasting

Forecasts for 24 hours ahead, with an outlook beyond that for a further 24 hours are termed **short-period forecasts**. Such forecasts have now achieved an accuracy of between 80 and 85 percent. Improvements in the forecasts depends on a variety of factors, including better data from satellites and improved meteorological understanding of the global weather machine. Above all, progress has been due to the performance of modern computers, the latest of which can perform 400 million calculations in a

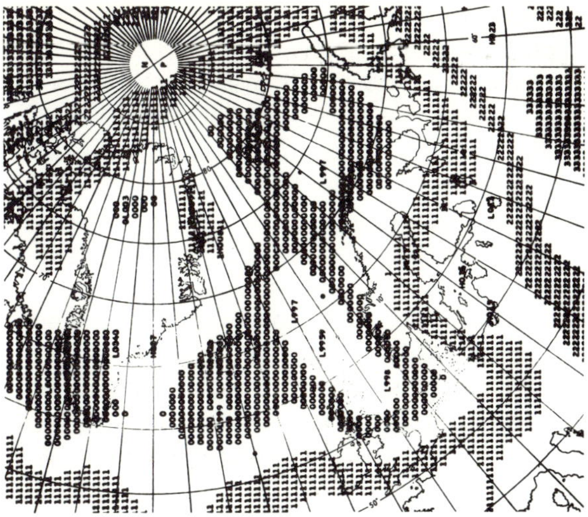

Figure 10.2 *Atmospheric patterns produced by computer*

second and produce a numerical forecast in chart form of the expected atmospheric patterns around the British Isles (see Figure 10.2). Weather radar is being used to produce a picture of where the rain is and its intensity and the forecaster can track cells of rain as they move across the country, and give accurate information of when and where rain can be expected in the next few hours. At present there is a network of five radar stations, and the images that are derived from them are sometimes used in the television forecasts shown by the BBC.

Medium-range forecasting

Forecasts for up to a week ahead involve about a billion calculations by the computer using world-wide data of the state of the atmosphere, oceans and land surfaces obtained from both ground-level observation and satellites. Even with huge computers the models of the weather are relatively coarse. For example, they calculate global weather on a grid of points 150 km apart. Even when the forecast is broadly correct, small errors can lead to appreciable local discrepancies. A small error in the timing of a weather system can mean that rain predicted for one day may actually arrive on the next day.

Even with computer and measurement advances the limit to useful forecasts is probably 10–14 days ahead. Beyond this all we can hope for is that we shall be able to identify what causes the global weather system to settle down into the abnormal patterns which lead to an exceptionally cold winter or a summer drought.

Long-period forecasting

For some years the Meteorological Office issued forecasts for the month ahead. They cannot be based on detailed calculations, but mainly by analogue. On the assumption that particular patterns and sequences of weather occur, data is fed into the computer to discover similarities between the weather patterns of the recent past and in former years. Where close similarities are found to exist, it is assumed that the weather will follow the same sequence, e.g. a relatively warm and wet January often follows a warm summer and dry, cold autumn.

Although there has been some success in the field of long-term forecasting, one of the problems is that it tends to be rather general. It can provide expectations for a region but actual figures and details for small areas cannot be scientifically calculated. These forecasts have now been terminated as an economy measure.

Study activities

1 Listen to or watch the short-range weather forecast on radio or television regularly for a week. Compare the forecast with the actual weather experienced in your area. How accurate do you think the forecast was? If there were inaccuracies do you think that these were due to: (a) poor timing of the arrival of precipitation, or (b) an unexpected development in the weather situation not foreseen by the forecaster?

2 If you are already familiar with weather charts and their interpretation, attempt your own forecast for 24 hours ahead, either by watching the television weather map with the sound turned down, or by looking at the weather maps published in many newspapers. How accurate were you?

Meteorological information services

Specialised information

Marine forecasts

The shipping forecast is broadcast on BBC Radio 4 (UK) on 1500m at 06.25, 13.55 and 17.50 hours, and then again at 00.30. The forecast is designed for commercial shipping and although it only lasts for about five minutes, it is able to convey a lot of information, which is given in the following form:

1 a gale warning for those sea areas (see Figure 10.3) where it is forecast the wind will equal or exceed gale force 8 in the next 24 hours;

2 the general synopsis, which is a brief description of the position of the major low and high pressure areas and their expected movement in the next 24 hours, together with mention of troughs, ridges, and fronts where appropriate;

3 the forecast for each of the sea areas shown in Figure 10.4. These are always given in the same order, beginning with Viking and ending with south-east Iceland. Consecutive areas may be grouped together if the forecast for each is expected to be generally the same.

4 Reports are given for a series of coastal stations around the British Isles. Normally these are recorded two hours before broadcast time so, for example, the 17.50 hours broadcast will be of the 16.00 hours reports.

A useful forecast for shipping close to land is the inshore waters forecast broadcast on BBC Radio 4 at 00.35 hours. This forecast is similar to the shipping forecast in form but, as the name suggests, some indication of local coastal variations is given.

The Meteorological Office also operates a ship-routing service, advising the captains of ships on the best course to sail to avoid hazards like gales or ice. The object of the service is to enable vessels to reach their destinations by the quickest and therefore most profitable route.

Figure 10.3 A shipping forecast

GENERAL SYNOPSIS at 1300 BST 16th APRIL

System	Present position	Movement	Forecast position	at
L 78	600 W IRELAND	NE sl	200 W BAIL	1300
FT	APP VALENTIA	st NE		
T	W SEA AREAS			

Gales	SEA AREA FORECAST	Wind (At first)	Wind (Later)	Weather	Visibility
	VIKING	V. 3	S 4/5		99
	FORTIES				
	CROMARTY				
	FORTH	S 3	5/6	occ •	m
	TYNE				
	DOGGER				
	FISHER	N 5	V 3/4		g
	GERMAN BIGHT				
	HUMBER	N 4/5 → S 3	4/5	occ •	g → m
	THAMES				
	DOVER	W 3	SW 4/5 (6)	occ •	m → p f pats l.
	WIGHT				
	PORTLAND	SW 4 → 5/6	(7)	occ •	m / p f pats.
	PLYMOUTH				
	BISCAY	SW 4 → 5/6		• / ,	m/ p f pats.
✓	FINISTERRE	SW 5-7 Loc 8		• → ∇	m / p → g
✓	SOLE				
	LUNDY				
	FASTNET	SW 4/5	(6-7)	• → ∇	m Loc p
	IRISH SEA				f pats at f.
✓	SHANNON	SW 5-7 Loc 8		• → ∇	m → g
✓	ROCKALL	S 5-7 Loc 8		• → ∇	m → g
	MALIN	S 4 → 5/6		occ •	m/g
	HEBRIDES				
✓	BAILEY	SE 5-7 Loc 8		occ •	m/g
	FAIR ISLE	SE 4 → 5/6		occ • l	m/g
	FAEROES				
✓	S.E. ICELAND	SE 5-7 Loc 8	E (4 in N at f.)		g

Mark gale areas Connect areas grouped in forecast

COASTAL REPORTS at 1600 BST	Wind Direction	Wind Force	Weather	Visibility	Pressure	Change
Tiree	S'E	3		11	10	
Sunburgh	SE'E	4	haze	2	12	
Bell Rock	SE'S	4		8	13	
Dowsing	SE	1		5	17	
Noordhinder	0			5	17	
Varne	SW	2		5	18	

COASTAL REPORTS	Wind Direction	Wind Force	Weather	Visibility	Pressure	Change
Ryal Sovereign	SW	4		11	18	
Ch. Ltvessel	SW'W	2	mist	2	18	
Scilly	SSW	4	,	4	16	
Valentia	S'W	4	•,	2	10	
Ronaldsway	SW'S	3	•	5	12	
Malin Head	SSE	3	•	8	10	
Jersey	SSW	3		8	19	

Figure 10.4 Sea areas included in the Daily Shipping Forecast

> **Study activity**
>
> Tape the broadcast of the shipping forecast, and then plot the information and analyse the results to produce a simple weather chart.

Land services

Meteorological information is also useful to farmers. For certain farming operations like crop spraying or haymaking, a more detailed forecast than is normally given may be required.

These detailed early-morning forecasts are broadcast by the London Weather Centre on the BBC Radio 4 *Farming Today* programme and the Sunday lunchtime forecast, *Weather for Farmers and Growers* on BBC1 television. In both cases, emphasis is given to the outlook period as this is important for planning work on farms. On the Sunday broadcast (and also on the Wednesday lunchtime one), the current weather situation over the North Atlantic and Western Europe is shown first, followed by the expected situations for mid-week. Special mention is made of any weather features which are likely to have a major impact on farming or horticulture, e.g. frosts, weather that allows the spread of plant disease, dry spells for haymaking, and so on.

The North Sea oil and gas industry requires accurate wind and wave forecasts, especially during times when rigs are being moved or wells sunk. The consumption of energy generally is influenced by weather conditions, especially temperature, so the gas and electricity producers need to know when there is likely to be increased demand due to a fall in temperature.

Manufacturing, building and construction industries all require advance knowledge of weather conditions, so that they can plan their production and distribution most effectively. Chapter 6 gives details of the significance of weather forecasting to the economic life of any developed country. The importance of shipping forecasts which are essential to primary industries like fishing, quite apart from transport and communications, has already been noted. Land communications are also affected by weather, but if given warning some precautions can be taken. For example, British Rail can run de-icing trains and activate heaters that are installed at points of key junctions, helping to reduce delays caused by freezing points and icing on lines.

General information

Radio

Weather forecasts are broadcast regularly on the various BBC services and details are given in the *Radio Times*. In addition, local broadcasting stations, both BBC and commercial, include weather information on their services. On the nationwide networks, Radios 1 and 2 have brief forecasts for the UK, with 'flash' warnings occurring on Radio 2 concerning snow, gales, fog and ice, of particular importance to motorists, and also flood warnings. Radio 3 transmits several brief forecasts, while Radio 4 has more detailed forecasts which include a statement on the general weather situation, a forecast for the next 24 hours, and an outlook for a further 24–48 hours. The complete broadcast is scheduled to take between three and four minutes and provides the most comprehensive general forecast available. At 00.15 hours, there is a report on the previous day's weather, the weather developments that are occurring, and the warmest and coldest, the sunniest and the wettest locations in the previous 24 hours, followed by a forecast for the following 24 hours.

Television

On BBC1, a forecaster of the Meteorological Office displays weather charts several times each day. The usual practice is to show the current synoptic chart for the European–Atlantic area, and this is then followed by one or more detailed forecast charts for the British Isles. These charts use a series of symbols to represent the likely weather over the country and the major symbols currently in use are shown in Figure 10.5. On certain BBC1 fore-

15	TEMPERATURE	Red figures on yellow background give positive temperatures in degrees centigrade. Black figures on a light blue background give freezing temperatures i.e. below zero centigrade
25	SUNSHINE	The yellow symbol represents the sun; the red figures in the centre show a temperature of 25 degrees centigrade
	CLOUD	A white cloud symbol indicates fine weather clouds that may be relatively thin and patchy
		A black cloud represents the thicker and more widespread clouds often associated with dull weather
	SUNNY INTERVALS	The sun symbol used in conjunction with a cloud in this way means some sunshine as well, particularly if the white cloud symbol is used
	RAIN	The dark blue teardrop symbols beneath the cloud indicate rain
	RAIN SHOWERS AND SUNNY INTERVALS	A combination of rain, cloud and sun represents sunny intervals and rain showers
	SNOW	The white snow symbols beneath the cloud indicate snow
	SLEET	The rain and snow symbols together beneath the cloud indicate sleet
	THUNDERSTORM	The symbol of a black cloud with a yellow flash represents the possibility of thunder and lightning
30 →	WIND SPEED AND DIRECTION	The black symbol represents the wind speed and direction, the speed printed in the centre in white is in miles per hour
	FOG	Fog is not represented by a special specific symbol; it is indicated by words on the map in general areas likely to be affected

Figure 10.5 Major symbols in use on television weather charts

Figure 10.6 Satellite photo of weather pattern over Britain and Euro

casts, the forecaster shows a recently received satellite photograph and draws attention to its main features over the British Isles.

On BBC2 weather forecasts, normally read by an announcer, are given after most of the news bulletins, while the majority of the independent television companies broadcast weather bulletins based on information supplied by the Meteorological Office, and their times can be obtained from the *TV Times*.

The press

Weather forecasts for all areas of the United Kingdom and for the Irish Sea and English Channel crossings are printed regularly by both national and regional newspapers. The unavoidable delays between the time of issue of these forecasts and the time of reading causes them to be less accurate than the up-to-date forecasts on radio and television. The most comprehensive weather coverage is to be found in the *Guardian*, *The Times* and the *Daily Telegraph*, with the first two including a chart of expected pressure distribution, while the latter contains the actual chart for the previous day at 12.00 hours. Weather information is also given for a wide spread of places over Europe and further afield, and during winter there is information about the Alpine ski resorts.

Weather information can also be obtained by telephone on an answer-phone service and by contacting the regional forecast centres. *Prestel*, the Post Office viewdata system, provides weather forecasts, as do the *Ceefax* and *Oracle* teletext pages.

Meteorological warnings

If severe weather conditions are expected in an area, general warnings may be put out between programmes on radio and television. These 'flash' warnings are issued if snow, gales, fog or heavy rain and consequent flooding are expected. In addition, during the winter warnings are given to drivers concerning road dangers, notably fog, snow or icy patches on roads. The warnings are particularly important where people need to take action to avoid the effect of the hazards. Coastal sailors and fishermen can take shelter after hearing a gale warning; drivers can avoid hilly routes if snow is forecast, and they may stay at home if fog is expected. If there is imminent danger of flooding, livestock may be moved from low-lying ground and furniture may even be moved upstairs. Sudden changes in the weather in mountainous regions cause deaths every year, but many lives are saved by those who take heed of the broadcast warnings.

Understanding and misunderstanding the forecast

Any forecast will only achieve its objective of communicating the views of the forecaster if it can be understood. Broadcast and televised forecasts have to be presented quickly and usually use a number of semi-technical terms or maps and charts. Research by psychologists has shown that a sizeable proportion of the population does not understand the terms or appreciate the message of the map. If, as so often happens, the listener or watcher is not giving his full attention to the forecast, he may well misunderstand. It is quite easy to hear the forecaster talk about rain tomorrow, but not to appreciate that he is talking about Scotland, when the listener happens to live in London. Another problem is that the forecaster has to divide the country into regions, and sometimes people cannot quickly relate where they live to the division being used.

Very often the forecaster must use terms like, 'it is probable that' or, 'some areas may have showers'. Some people find such terms vague and would prefer a numerical percentage to express the confidence felt by the forecaster. In parts of the USA it is common to hear phrases like, 'there is an 80 percent chance of rain tomorrow'. In Britain few people realise the accuracy of the 24-hour short-range forecast. The forecaster continues to be doubted and the times when he has been wrong are remembered far better than the many times he has been right!

Study activities

1 Make a list of the ways in which you personally might benefit from good weather forecasts. Could you make better use of them than you have in the past?

2 In a number of major cities, including London, Glasgow, Southampton and Newcastle, weather centres have been opened. Try to visit such a centre. Does it have a window display—showing current or forecast weather? There may be leaflets or booklets for sale inside, describing the work of the Meteorological Office.

3 Compare the presentation of the weather forecast by the different television channels. Which do you think is the most effective and easily understood forecast? Why is it superior to the others?

The Meteorological Office in the UK employs 2500 people, the majority at its headquarters at Bracknell in Berkshire. It operates a network of over 100 stations all over the country which serve the community by observing, investigating and forecasting the ever-changing patterns of our weather. Many people's livelihoods and sometimes even their lives depend on the elements, and at a cost to the taxpayer of a little over £1.00 per year (in 1985), our meteorological service must be reckoned to be good value. The television 'weatherman' is familiar to us all, but he is only at the tip of the iceberg, and the visible representative of the world-wide organisation that is necessary to forecast the weather. Although the general weather forecasts for the public meet the needs of most people, specialist customers require precise services.

11

Atmospheric energy resources

Frequent reference is made to an **energy crisis**, in that fuels like coal and oil will become even scarcer and more expensive in the future. At the same time the presence of wind and watermills in the landscape are a reminder that there are alternatives to the use of fossil fuels and that man has exploited the atmosphere, harnessing solar and wind energy for many centuries. Perhaps it is not surprising that there is now more interest in alternative energy sources than for a long time, and pressure groups (like Friends of the Earth) draw attention to safe, non-polluting sources of energy which also have the advantage that they are renewable and will not be exhausted in the future. Research and development of these energy sources is taking place at present, and each of them will now be examined.

Wind power

The first consideration in any proposed wind-power scheme must be the amount of power which is available at the proposed site. This is related to wind speeds in the locality and the total amount of energy available is a function of the number of days for which the wind blows at a particular speed. Wind power varies as the cube of the wind speed; the wind does not blow constantly, and even in windy locations there will be days with light winds or calm conditions when the wind is not available to generate electricity. Because it is difficult and expensive to store electricity a constant supply of power cannot be relied on. These problems can be partly overcome by using very large wind generators and siting them very carefully. For domestic purposes wind power is most likely to be used in very isolated locations, such as small islands cut off from the national grid supply.

Even though wind generators can look elegant, few people would wish to see them built in large numbers in beautiful locations, although perhaps they could be sited in offshore locations. The largest generator to be built in the UK so far is the 22 metre high installation at Burgar Hill in the Orkneys, (see Figure 11.1) which is producing sufficient electricity to supply 100 homes. Research shows that for more than 70 percent of the year wind speeds exceed the 5 metres per second required to spin the turbine's rotor-blades. A prototype of a

Figure 11.1 The 300 kw aerogenerator at Burgar Hill in the Orkneys

vertical-axis wind machine, which is thought to be more efficient than other types of generators, is being constructed at Carmarthen Bay, south-west Wales, due for completion in 1986. In the United States more than 2000 microprocessor-controlled windmills have been erected at the Altamont Pass in California, where in summer cool air from the coast is sucked along the natural funnel of the pass and the wind blows consistently enough to provide commercially viable wind power. The electricity generated is sufficient for the needs of about 5000 homes in the area. In Europe, Sweden is an example of a country that wishes to phase out its nuclear power plants and cut its dependence on imported oil—two large aerogenerators have been built, one on the island of Gotland and one on the mainland.

Power from the waves is dependent on the wind and this could make a contribution to energy needs, but again, if wave power were the only form of energy available, its variability from day to day and season to season would be a problem. Compared with wind power, wave power has a high availability—there are always waves arriving at the coast generated by storms often in the North Atlantic. In designing a wave-energy machine, size and cost are important and experiments are proceeding to evaluate possible equipment. In the Irish Sea, storm waves of 400 metres length between crests with a sixteen-second period occur and would have a maximum power content of 700 kw per metre. While there is immense power available from the waves, it is not necessarily economic to tap it, and to date (1985) the government has cut funding on further research into the use of wave power.

Solar energy

The amount of solar radiation at any location depends on the latitude, the season of the year and cloudiness. In order to cope with periods of cloudy days a building heated by solar energy requires a standby heat source that uses conventional fuels. Even in Britain's notorious climate, an average of 1400 hours of sunshine a year will deliver some 900 kw hours to each square metre of England and Wales. A black surface will absorb most of this energy, turn it into heat and trap this heat beneath a layer of grass or plastic with water pipes attached to them so that it can be carried into the building to provide hot water. A south-facing roof is the best place to fit a solar panel and it has been calculated that annual fuel bills can be reduced by 20–24 percent after installation of a solar heating system. About 25 000 home solar units have been installed in Britain. At Milton Keynes, a 'solar house' has been built with a large sloping roof, to maximise the sun's energy. In sunnier climates, for example in Mediterranean countries and in the Middle East, solar energy can provide a more regular and consistent source of power. Because the amount of solar radiation available is at a maximum in the summer, this energy can offer a cheap way of heating the water in a swimming pool and this technique is used at Eton College. In the clear air of the Pyrenees the French are experimenting with a large number of solar collectors placed close together. The Themis solar-energy power station near Torgasonne is the largest such installation in Europe and will produce about 2.5 MW of power.

Work has started in Sweden on one of the world's largest energy-saving projects: water

Figure 11.2 A purpose-built 'solar house' in Milton Keynes, Buckinghamshire. The roof is made up of solar panels

heated by solar energy during the summer months will be retained for use in winter in an underground aquifer. It will be recovered in winter and have its temperature boosted before being fed into central-heating networks. The favourable cost estimates suggest that subterranean heat-store construction programmes will be expanded in the future.

Conclusion

Atmospheric energy resources offer clean energy with no pollution, independent of overseas sources of fuel supply. The raw material is limitless, and as fuels like coal and oil become scarce and more expensive, more money is likely to be spent on perfecting the use of these energy resources. The exploitation of these energy resources is now showing commercial promise, and as more experience is acquired they seem likely to become increasingly important as alternative sources of energy.

Study activities

1 Try to visit the Centre for Alternative Technology at Machynlleth, Powys, in mid-Wales, where alternative sources of power such as wind and solar power are being exploited. Visits can be made to the Centre by individuals and school groups. (Open 10 a.m.–6 p.m. Tel. 0654-2400).

2 Radio and television programmes and newspaper and magazine articles are often devoted to a consideration of these atmospheric energy resources. Try and listen to or watch such programmes and keep a scrapbook of press articles.

3 Ask your family how much the annual fuel bill is for your home. A 5 metre square solar collector costs about £1500 (at 1985 prices). If such a collector can save 25 percent of the annual fuel bill, how long will it take for the collector to pay for itself? Now repeat the calculation, assuming that the annual fuel bill increases by 5 percent per year.

12

Controlling the weather

It was Mark Twain who said, 'Everybody talks about the weather but nobody does anything about it.' This comment ignores the fact that since the dawn of civilisation, people have been trying to increase or reduce precipitation, or to alter the pattern of the winds. Numerous methods have been tried—ranging from ringing the church bells to firing rockets at clouds—while still today primitive tribes engage in rain dances.

In Britain, it was appreciated for many centuries that it was possible at least partly to control the climate: for example, walled gardens providing a warmer and less windy climate allowed the growth of such fruits as peaches and grapes.

Today's farmer continues these attempts at climate control through the use of irrigation equipment. In south-east England, higher crop yields can be achieved in nine years out of ten by the use of irrigation, although before installing the equipment, the farmer has to estimate if the increased revenue to be received from higher crop yields will offset the capital expenditure. Fruit growers can take practical action to minimise late frost damage once the trees are planted, by lighting bonfires. The smoke produced can reduce heat loss on cold nights and, hopefully, avoid a drop in temperature sufficient to produce frost.

The degree to which people can change climate generally depends on the scale of the attempt. The key to any climatic change is an alteration of the heat and water balance of the system involved.

Why attempt to change the climate?

In 1942 a German meteorologist seeded a small cloud over Prague with a silica dust, with the result that a mass of snow crystals formed. Since then, weather modification has moved from the realms of science fiction to reality and to scientific respectability. In the United States and in many other countries, programmes to increase precipitation by seeding suitable clouds are regularly carried out. By this means rainfall can be increased by 10–15 percent. Providing additional rainfall at a critical time can increase agricultural productivity. As a result many commercial firms have sprung up to 'make rain'. Desirable as weather modifications might be, they raise a host of economic, moral and even political questions. One farmer may want extra rain to fall on his land, but his neighbour may not—imagine the litigation that might occur from crop failure or damage due to these attempts.

Experts have argued whether a flood which occurred in Rapid City, South Dakota in January 1972 was caused by cloud-seeding experiments. Conflicts of interest resulting from an expansion of weather modification activity will raise very fundamental questions like: Who owns the atmosphere? Can individual nations lay claims to part of it? The two superpowers, the USSR and the USA, have reached preliminary agreement to ban weather modification as a weapon of war in the future. Attempts to make use of weather modification techniques had been used by the Americans in the Vietnam War in the 1970s. The military authorities seeded clouds to produce rain in order to hinder the enemy's ability to fight.

A further basic motivation for modifying the weather, for more honourable reasons, is to reduce the economic and social losses that result from severe weather events. Hail can damage crops, lightning starts fires, and because violent hurricanes cause widespread damage, experiments have started to try to control them. So not only is weather modification considered on the scale of improving profits, but on the more fundamental level of protecting lives and staving off famine. However, although much has been learned about the basic mechanisms of the atmosphere and about means of altering these processes, progress has been relatively slow compared with other fields of scientific research.

Can the climate be changed for everyone?

Is it possible to improve the climate of large areas or perhaps to offset a predicted deterioration of climate by large-scale, climate-control schemes? Among the grander proposals for climate modification are the dusting of ice and snow with carbon black to speed melting and the creation of dark areas on land, like large forests, to produce convective rainfall. We have the technological ability to alter climate deliberately and the future physical influence of mankind on the climate environment could be very significant. Some possible schemes that could be proposed to modify or control the climate are illustrated in Figure 12.1. Do they make sense? The atmosphere is a highly complex and interactive resource common to all nations. Who can decide and implement climatic modification schemes?

The layer of drifting ice that covers much of the Arctic Ocean varies in thickness from 2 to 3 metres. Spreading black particles, such as soot, would reduce the reflectivity (albedo) and could cause the ice to disappear within three years. An open ocean would produce a more moderate climate for the high latitudes, but what would happen to sea levels? Another proposal involves damming the Bering Straits, then pumping water from the Arctic into the Pacific while inducing warm Atlantic water to

Figure 12.1 Engineering schemes that could be proposed to modify or control the climate

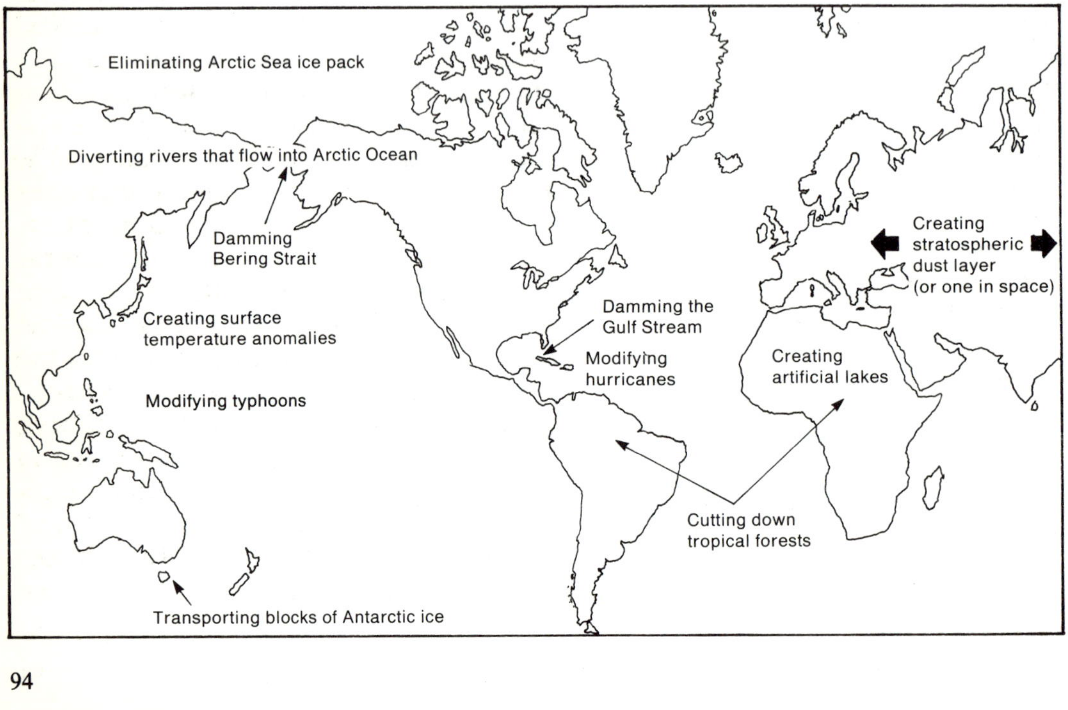

flow into the eastern Arctic. This again has a superficial attractiveness. However, with both schemes, a world-wide rise in sea level could well occur, therefore inundating large areas of low-lying coastal regions.

Hurricanes might be steered by pumping cold water to the surface ahead of them, thereby reducing their supply of energy. One problem would be to decide who would have ultimate authority for steering the hurricane, and who would pay compensation if the track of the storm were displaced.

It is not easy to produce effective legislation relating to weather modification, and its future is unclear. While local attempts at increasing rainfall may be acceptable, more major projects will require a large amount of study to reveal whether the disadvantages of implementing them outweigh the advantages. Without high-level discussions and decisions, Mark Twain's comment could be reworded: 'Everybody is doing something about the weather, but nobody is talking about it.'

Study activities

1 Write an essay discussing the problems and prospects of purposeful weather modifications, including the ethics involved.

2 Try to visit a walled garden (there are many examples in the gardens of English stately homes open to the public). List the crops grown in the garden and try to discover which could not be grown without the protection of the enclosure.

3 (This activity requires a small group of students to work together.) Imagine that a commercial cloud-seeding company has been approached by a farmer in the mid-west of the USA to produce additional rain over his drought-stricken crops. Neighbouring farmers growing different crops are about to start harvesting operations and are hoping that fine weather will continue. In the local small town there are a group of environmentally conscious citizens who feel that it is wrong to meddle with the weather and introduce chemicals into the atmosphere. Other townspeople fear that if the rain-making operation is too successful, the local tourist industry will be damaging, relying as it does on the attractions of a local lake which is a popular spot for swimming and sunbathing.

Decide who in the group is going to represent:

(a) the 'rain-making' company;

(b) the farmer who wants more rain for his crops;

(c) a farmer about to commence harvesting;

(d) a spokesperson for the local environmental group;

(e) an official of the local tourist office.

Prepare a short contribution to be given at a meeting to discuss the controversy that has arisen in the community over the rain-making plans. Meet together as a group and present your views. Who do you think has the strongest arguments?

Conclusion

The relationship between people and their environment is a central issue in geography. Weather and climate are important parts of the environment and in this book we have seen that the response of individuals and groups to their atmospheric environment is varied and often complex. While the climate system has an impact on society and its activities, people also have an impact on the climate system. The earth's atmosphere is a resource and like any resource it can be used or misused by mankind. The key to managing climatic resources is the effective use of information. If we had perfect weather forecasts we could adjust our activities to take advantage of the weather with relative ease. Although our forecasting ability is imperfect climatic information can still be valuable if used effectively. At all scales people seek to achieve a better adjustment to the climatic environment. We often think of weather as 'good' or 'bad' when in reality it is neutral and the key to progress lies in our own attitudes and actions.

Bibliography

Modern Meteorology and Climatology, Chandler, T. J., Nelson, 1972
Worldwide Issues in Geography, Hart, C. (Ed.) Collins Educational, 1985
The Value of the Weather, (2nd edition) Maunder, W. J., Methuen, 1986
Environmental Hazards in the British Isles, Perry, A. H., George Allen & Unwin, 1981
Environmental Systems, White, I.D., Mottershead, D. N., and Harrison, S. J., George Allen & Unwin, 1984
Physical Geography—flows, cycles, systems and change, Wilcock, D., Blackie, 1983
Meteorology, Wright, D., Basil Blackwell, 1983

Index